W9-CNP-214

CATS'
KINGDOM

CATS' KINGDOM

JEREMY ANGEL

WARNER BOOKS

A Warner Communications Company

The photograph on page 5 and the bottom photograph on page 81 were taken by Chiyoko Angel.
All the rest were taken by the author.

Warner Books Edition

This Warner Books edition is published by arrangement with
Souvenir Press, Ltd., 43 Great Russell Street, London WC1B 3PA

Warner Books, Inc., 666 Fifth Avenue, New York, NY 10103

Manufactured in Spain

First Warner Books Printing: November 1987

10 9 8 7 6 5 4 3 2 1

Library of Congress Cataloging-in-Publication Data

Angel, Jeremy.
Cats' kingdom.

1. Cats—Behavior. 2. Cats—Japan—Hokkaido—Anecdotes. 3. Cats—Pictorial works. 4. Mutsugoro's Animal Kingdom (Japan) 5. Angel, Jeremy. 6. Hata, Masanori—Homes and haunts—Japan—Hokkaido I. Title.
SF446.5.A54 1987 636.8 87-40172
ISBN 0-446-51367-9

Designed by Giorgetta Bell McRee

CONTENTS

1

THE BIRTH OF THE KINGDOM

Eastern Hokkaido, in the northern part of the Japanese archipelago, is not the kind of place in which most people, including myself, would choose to settle permanently. The summers are short, the winters long and bitterly cold, and most of the landscape is a flat alluvial plain across which northwesterlies from Siberia often sweep with numbing ferocity. But to a young writer and onetime zoologist, Masanori Hata, alias Mutsugoro, it seemed ideal. In 1968, having just won a literary prize, he was looking for a place where he could live surrounded by nature and all the animals he could support. Eastern Hokkaido was remote, still rich in unspoiled countryside, sparsely populated by dairy farmers, horse breeders, foresters, fishermen and seaweed collectors

—people who lived close to the earth, in tune with its rhythms, just as he wished to do. Land was plentiful and, by Japanese standards, very cheap. In 1969, when he was thirty-four years old, Masanori Hata moved from Tokyo to this region, with his wife and daughter, and has been there ever since.

He first settled on a tiny uninhabited island, no more than a square mile in area, just off the Pacific Coast, moving into a shack with his family, his dog and a brown bear cub. After a year, however, he was forced to find more spacious accommodations. Donbe, the bear, was growing too big to be allowed to roam free any longer, and other animals had joined the household—four dogs, two cats, three Hokkaido ponies and two tame crows. He leased and fenced in twenty acres of land just along the coast from the island and called the place Mutsugoro's Animal Kingdom. This was a somewhat grandiose name for what was little more than a motley array of paddocks, enclosures, hutches and stables designed to house his private menagerie, but none seemed more appropriate for his vision of the place: a sanctuary where he and other people of his choosing could lead a relatively free, if hard, existence in the company of animals, learning about them and, through them, about themselves.

Over the years, as animals were brought to him for care and shelter and stayed to become permanent residents, the menagerie grew, a mixture of domestic and wild animals. Each new arrival, in its history, growth and relationships both with its fellow animals and with its human keepers, provided more material for an unending stream of books from the prolific pen of Mutsu-san (as everyone in the Kingdom called him). The royalties that resulted just about covered all the expenses involved in keeping so much livestock, but the Kingdom was soon so famous that Mutsu-san could have made life much easier for himself by opening it to the general public, thousands of whom were willing to trek up to that remote region to catch a glimpse of the animals immortalized in his books. However, he steadfastly rejected any suggestions to this effect, feeling that the place would lose its unique character, and he himself a lot of his freedom, by such a move.

He did, of course, need help in the actual management of the place and the daily care of the animals. Early on, he had been joined by his brother Hige, a photographer, and the latter's wife, and in the years that followed he recruited a succession of young people from the ranks of his readers, who he thought had something to offer the Kingdom and something to gain in return. Many left after a few days or weeks, unable to stay the course; of those who remained, most left after a year or two to follow a career, but some stayed longer, and two of them, now married, have been there for thirteen years.

I first heard about Mutsu-san and his activities in 1976, when I was in Hong Kong, from a Japanese friend who was a fan of his. I was single, carefree, irresponsible, engaged in work for which I had little aptitude or enthusiasm, and was looking for any excuse to move on. I had grown up, at least from the age of eight, in the English countryside, with dogs and cats in addition to two brothers as my childhood companions, and I was drawn to nature and animals. I was also an inveterate map-gazer, perhaps as a result of my early upbringing in the tow of my father, an itinerant civil engineer. By the time my family eventually settled in England, I had passed two years of my life in India and a further two in Canada, and had left my footprints on other bits and pieces of North America, Asia and Europe —wanderings which no doubt helped to instill in me a deep fascination for faraway places and exotic peoples.

As I listened to my friend's account of Mutsugoro's Animal Kingdom gleaned from his books, I realized that here was a grand opportunity to indulge my passions. You couldn't, if you were an Englishman, find too many places further away than Hokkaido and, judging from what I heard, Mutsu-san seemed to be a pretty exotic person in his own right. Then there were the animals and what appeared, from the map of Japan that I had dug out, to be large tracts of mountains, open countryside and stretches of desolate coastline—a far cry from noisy, smelly, humid, claustrophobic Hong Kong. I had grown fat on too much delicious Cantonese cooking and too little exercise; Hokkaido and the Animal Kingdom promised clean air, physical labor and stimulation.

I got my friend to write a letter of introduction to Mutsu-san on my behalf, considering, in my vanity, that I might be useful to him in some way. It was a long shot, but I had nothing to lose. Once the letter had been sent, I put it completely out of my mind and so was absolutely astounded when, two weeks later, a reply arrived, asking me to come anytime, to stay as long as I liked and, modestly, to expect very little. While I could hardly believe my luck, it seemed somehow fitting that, after all my failures at securing or remaining in more humdrum employment, my savior should turn out to be a diminutive literary genius—as my friend described him—holing out in the middle of nowhere on the other side of the world.

In July of that year, after making a three-week trip around China (the only really good thing to come out of my stay in Hong Kong), I boarded a plane to Tokyo and headed north almost immediately, but at a leisurely pace, by local trains, buses and hitched rides, in order to get a taste of the country and its people. My bag was loaded with twenty-year-old U.S. Army Japanese-language texts, a pair of size 12 Wellington boots purchased in Peking (almost impossible to obtain in Japan, where people tend to gasp at the sight of feet as large as mine) and a very thickly padded Mao jacket to get me through the subzero Hokkaido winter, should I find the place agreeable enough to stay that long.

As things turned out, although life in the Kingdom was not always easy, and although I quarreled with Mutsu-san at times, I liked the place, the man and his human and animal family enough to spend not one, but seven, winters there. During that time I learned to speak and read Japanese reasonably well, and even to write enough to be able to pen the Japanese version of this book largely by myself. Under the tutelage of Hige, I also discovered the joys and frustrations of photography, an art I find every bit as demanding as writing but far more pleasurable. I spent endless enjoyable hours in the company of the animals, playing with them, sleeping with them, feeding them, riding them, tending their illnesses and injuries and just plain watching them. They helped me to recall some of the zoology I had learned at Oxford, and to bring it vividly alive, but they did far more than this. Through intimate daily contact with them, I rediscovered the "magic," for want of a better word, of animals. For some years before arriving in the Kingdom, I had lived quite happily without the presence of even a cat or a dog. Now I find myself again in a similar situation, but after seven years in the Kingdom, an animal-less existence is no longer nearly so bearable, and I live and work for the day when I can once more at least walk the countryside with a dog, and snooze with a cat on my lap.

Once I had become sufficiently proficient in Japanese and with my camera, I found that I was able to be useful to Mutsu-san, and I traveled with him as his photographer/interpreter/companion throughout Australia, Europe, Sri Lanka and southern Africa, experiencing some of the most memorable adventures of my life so far. And I fell in love with Chiyoko, who looked after the Kingdom's pack of dogs, and married her in 1980. (She had had much more difficulty than I in being accepted into the Kingdom, having been rejected twice. Finally she wrote to Mutsu-san, describing her work on a dairy farm; her account of cleaning the cows' backsides and of the insights she gained thereby into their individual characters so tickled Mutsu-san's sense of humor that he invited her to join him.)

More than anything else in the Kingdom, however, what occupied my working hours was the cattery.

The idea of a special house for the cats was first broached one evening in June 1978 when we were all sitting round the long dining table in the living room, for the first staff meeting since Mutsu-san and I had returned from Sri Lanka. We had been away for over three months, researching for a book on working elephants, and what with recounting our adventures and catching up on all that had happened during our absence, it was already fairly late when Mutsu-san suddenly brought up the subject.

"The cats would appear to have been as busy as ever while I've been away," he observed, casting his eyes down the table toward the other end of the room, where six of the animals were enjoying a rare chance to monopolize the sofas. "How many do we have now, Hiroko?"

Hiroko, the girl responsible at that time for feeding and caring for the cats, raised her head to the ceiling and began counting on her fingers. "Thirty-three," she replied eventually.

Yes, thirty-three cats. About a dozen of

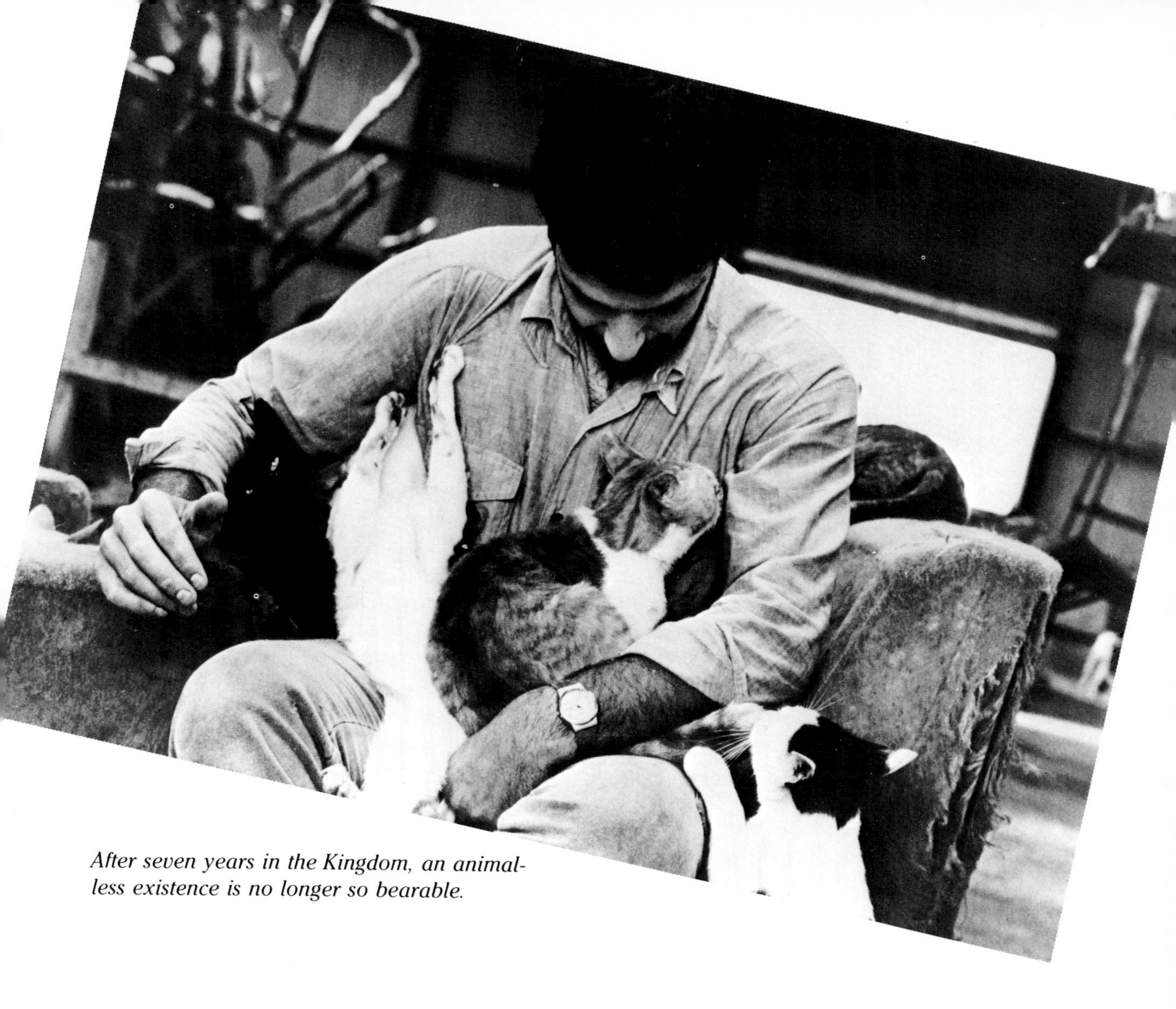

After seven years in the Kingdom, an animal-less existence is no longer so bearable.

these were immigrants which had either been dumped in front of the Kingdom gate or been taken in at the request of friends and neighbors whom it was impossible to refuse. The remainder, of course, were the offspring of these cats, and the offspring of those offspring. While the number of mature cats was still small, we had been able to find homes for most of the kittens born, but gradually production had outpaced demand, and suddenly, within the space of that spring, it seemed that we were swamped with cats. By sheer weight of numbers, they were beginning to make the house—especially the television room where they were most heavily concentrated—look like the set for a feline version of Hitchcock's film *The Birds*.

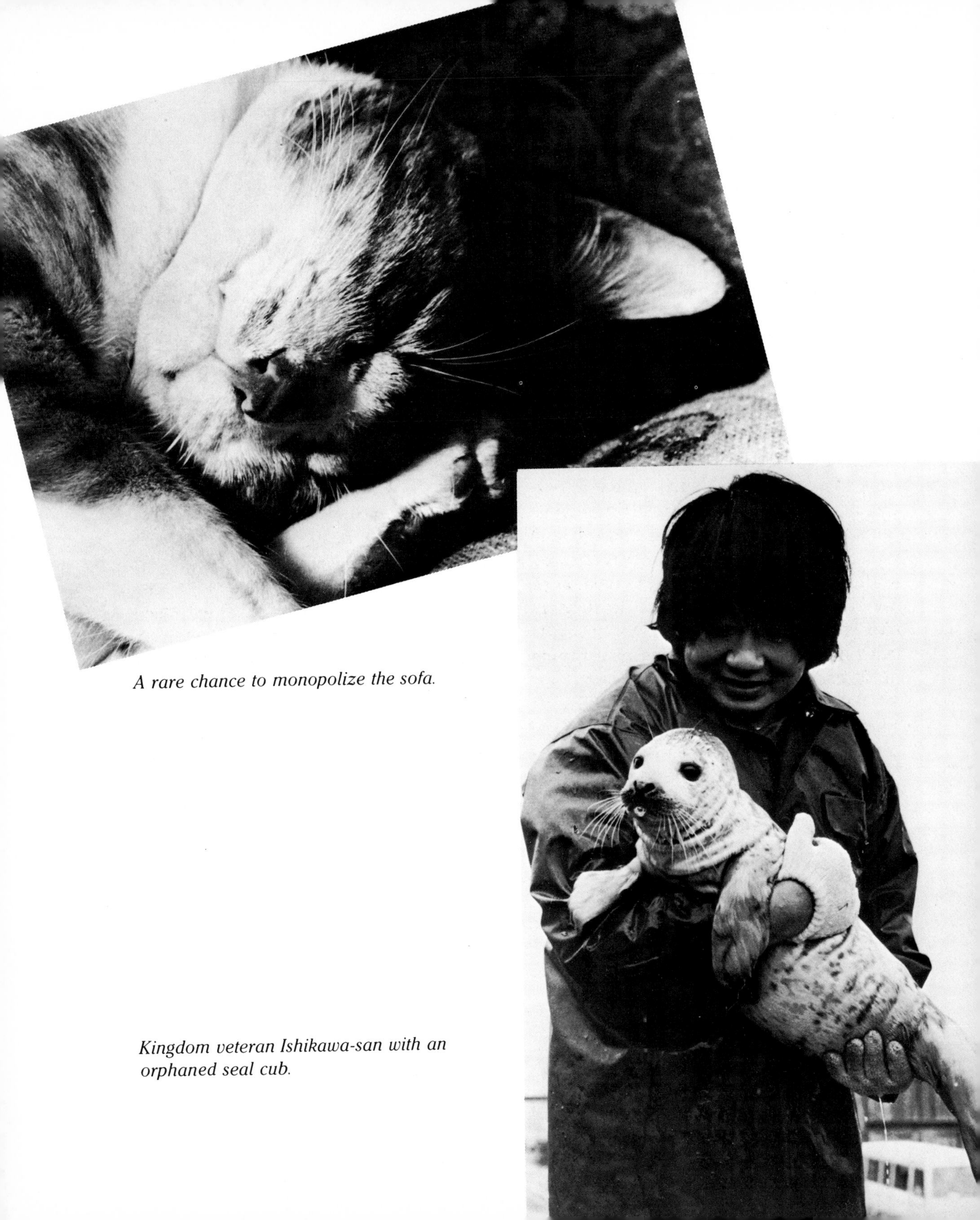

A rare chance to monopolize the sofa.

Kingdom veteran Ishikawa-san with an orphaned seal cub.

The television room was where Mo, a big blue tom with a stubby tail who, although only six years old, was the senior cat of the household, lorded it over a harem of queens and their juvenile offspring—a grand total of sixteen cats. Although Mo was a perfect gentleman toward his ladies, and almost never raised a paw against the hordes of kittens who seemed to regard him as an ideal subject for their horseplay, he did not lightly suffer rivals. Any of his male progeny who had reached an age and size that he considered threatening to his authority would be bullied with such spite and persistence that we were forced to remove them from his reach. Waru had gone, and then Sabu, Nibu and Panda, and by the look of things, Kenjiro would be the next. Poor Kenjiro! One could not hope to meet a soppier young tom. He would have been the last one to harbor ambitions of usurping his father's throne, yet during the previous few days he had become the target for Mo's aggression. He was now spending most of his time on top of the wardrobe in the corner of the room, staring fearfully down in Mo's direction. He slept there, ate his meals there, and even made one corner his lavatory—and any cat that turns its own sleeping quarters into a lavatory is a very desperate creature indeed.

The living room served as an asylum for such exiles, and was at that time occupied by the young toms Sabu, Nibu and Panda. Waru, the most pugnacious of Mo's offspring, had also passed some time there, but, seemingly determined to make up for his expulsion from the television room by setting himself up as dictator of the living room, he had made life so miserable for the other cats that Chiyoko had removed him to her own room, along with a couple of kittens who, she hoped, would teach him a little tolerance while keeping him company.

Also resident in the living room, because she was everyone's favorite, was Tobo, a Siamese queen who had been picked up close to death along the back road to the Kingdom one winter night. Making up the half dozen present that evening were two kittens who were there merely because kittens are fun to have around.

The stairs and upstairs landing and corridor served as home for the remainder of the cats, four toms and four queens, who got on with each other surprisingly amicably. However, sleep for the human inhabitants of the upper rooms was not always easy with those eight cats thumping up and down the corridor at all hours of the night, whenever the mood for play overtook them. One upstairs resident, Mutsu-san's mother, Granny Tokie, was finding the situation increasingly unbearable. The only person in the house with an aversion to cats, she was, under the circumstances, remarkably long-suffering, but the upstairs cats found themselves becoming more and more often the target of stinging swipes from the flyswatter that Granny brandished on her frequent nocturnal trips to the lavatory. The noise created by such violence was guaranteed to awaken anyone whom the cats had not already disturbed.

I must admit that, when I first arrived at

Once the cattery was established, we made a practice of introducing any new puppies to the cats. These two will have learned a lifelong respect for cats.

the Kingdom in 1976, I was surprised and disappointed to find that the dogs were not allowed into, or the cats out of, the house. Where the dogs were concerned, I could appreciate the impracticability of having the Kingdom's twenty-odd hounds, many of them large, lying around an already crowded household. The dogs did not lack for company, had plenty of shelter in their kennels and the stables, and seemed to flourish in their freedom. But why weren't the cats—about ten of them at that time—given the same freedom? To my mind, the only really happy cat is a free cat, one that is able to wander out at will, to climb trees or step soundlessly through long grass, to hunt mice and rats, to sunbathe on roofs, to seek the solitude that its nature demands. The sight of the cats gazing out of the windows, at times merely pensively, at other times with their eyes fixed longingly on sparrows by day and moths by night, saddened me considerably.

As soon as I could make myself understood, I asked the reason for this unhappy situation and was told the story of Ma, the Siamese tom who had been the founding father of the Kingdom's cat population. He had been free to come and go as he pleased, and used that freedom to court all the females in the vicinity. Mo is in fact one of the offspring resulting from Ma's activities, according to the fisherman who brought him to the Kingdom as a kitten. But Ma's life as the local feline Lothario was cut short early one morning by a murderous attack by some of the younger dogs. From that day on, any cats who came to live in the Kingdom had been confined to the house.

This sad tale only added to my concern, for it was clear proof of a lack of foresight in raising the dogs. Almost any puppy raised with a cat will come to regard it as a friend, or will at least learn, through the painful experience of sharp claws and fangs, a lifelong respect for cats. If only a positive effort had been made to educate the Kingdom's dogs at puppy stage, this segregation could have been avoided.

However, the damage was done, and the cats were obliged to pass their lives in the increasingly crowded conditions of the main house, to the great inconvenience of the human inmates.

"Thirty-three cats, eh? That's no mean total," mused Mutsu-san that evening. "Ah well, the more cats there are, the more they can teach us, not only about themselves but also about *our*selves and about animals in general."

"That's all very well, but you don't have to do all the cleaning every day," lamented Mutsu-san's wife, known to everyone as Obachan, or "Auntie." "And if the present situation is tough on us, it's even tougher on the cats. For all our sakes, let's operate on them and at least prevent their numbers from growing any further."

"Oh no, unthinkable!" replied Mutsu-san. "All this potential wasted? The more the merrier is what I think."

"Then you're proposing that we simply hand over the whole house to the cats," declared his wife, "for that will be the result if we don't do something quickly."

"No, no. Rather, I propose that we give the cats their own house. Let me explain what I have in mind."

Standing in a row along the rise above the Kingdom drive were four vast corrugated iron sheds that Mutsu-san had built at a time when he was feeling particularly flush. Each had an area of over four hundred square yards, with a concrete floor and windows already fitted. The first two were divided into pens where the sheep and goats were confined at night, indoor quarters for the foxes and others for sick animals, a pool for orphaned seals (two or three of which were brought to us each year for care until they were big enough to be returned to the ocean), and so on. But the end two sheds were being used for nothing more than storage of old furniture, machinery and other junk that had accumulated over the years.

Cats are far more social in their habits than most people suppose.

Mutsu-san proposed that we convert those two sheds into a single, huge house for the cats by joining them together with a passage, catproofing the windows and doors, filling them with old furniture, tires and logs, fitting aerial walkways, sleeping boxes and so on. All the cats should then be set free there. He reckoned that the two sheds combined could accommodate a hundred, even a hundred and fifty cats, with ease and, irrespective of numbers, each cat would have many times the space available in the main house in which to roam. They would be able to exercise and play, to love and to fight, to interact in any way that they wished on a scale that was not feasible under their present circumstances.

"To be sure, the toms, and probably the queens, too, will fight at first. There may even be deaths from such fights — we'll just have to see. But I don't think that violence will reign for longer than the first few days. Although the living area will be much larger, it will still be an enclosed environment. The toms will have to come to some arrangement about coexisting in that area, and it's such arrangements that interest me.

"Cats are generally considered pretty antisocial creatures. I know they can't be considered social animals in the same way as dogs or lions or sheep or ants, but I've lived with cats now for years, and I reckon that they are a lot more social than most people suppose. Yet remarkably little is known about their social behavior. In those sheds, left to do as they please, the cats will be forced to socialize to the limits of their ability, and in so doing will give us an ideal opportunity to observe the extent to which they can live with each other, and the ways in which they achieve such coexistence."

And that is how the cattery came into being. It was understood, right from the start, that the realization of the project was up to any of us who wanted to take it on, and it was I who volunteered. I was as curious as Mutsu-san to find out exactly what would happen once all the cats were released into those two sheds and left to do as they liked. Cats had fascinated me from an early age, and the project also seemed an ideal photographic challenge.

But I was also concerned for the welfare of the cats. I could foresee all sorts of problems with Mutsu-san's proposal, not least of which would be overpopulation again in two or three years' time; the cats would continue to breed, in ever greater numbers, and so the sheds were no final solution. Moreover, the cats would be "inside," not "outside" where I wanted them to be. But I had had enough of seeing them making do with a few square yards of living space. In those sheds they would have several hundred square yards, and I felt that they could not help but be happier creatures as a result. It was for this reason, as much as for scientific discovery or photographic ambition, that I decided, the moment that I heard Mutsu-san's plan, to make it my business to bring the cattery into existence.

2

A BLOOD AND THUNDER START

September 1, 1978, was day one of the cattery. During the morning, Chiyoko and I had released the cats into the sheds, batch after batch, but anyone ignorant of this fact would have been hard put to believe, on peeping into the cattery at noon, that there were indeed more than thirty cats inside the place. A preliminary scan would have revealed only two or three, hugging the walls as they skulked from one dark corner in search of what they hoped would be a darker one. Otherwise, virtually the only movement in the place was that of the cats' eyes, peering anxiously from assorted hideouts to take in their surroundings. The silence was intense.

More than two months had passed since that night when

Mutsu-san had first broached his ideas on the creation of the cattery. I had set to work immediately and had got the sheds ready for habitation in just over a month, but then another trip to Sri Lanka had come up and I had had no choice but to put off transferring the cats until after my return. I had regretted the delay, as I had wanted to give the cats, until then accustomed to the luxury of central heating, as much time as possible to adapt to their new environment before the onset of winter. As it was, although the days were still warm, the forests were already beginning to take on the hues of autumn, and subzero nighttime temperatures were now only a couple of months away. If the cats were to be given a chance to grow thick winter coats, every day counted, and so we had transferred them from the main house on the morning after my return.

I knew that the cats would need time to appreciate my efforts at furnishing their new quarters, and their initial reaction of fear was totally expected. Even cats which have not led the confined lives of those in the Kingdom resent very much being suddenly deposited in strange surroundings and, predictably, our cats felt far from safe. They suddenly found themselves in a world which, while equipped with a roof, was far larger than their previous one and was filled not only with unfamiliar objects but with a host of unfamiliar faces. They were justifiably terrified.

However, starting with the juveniles, one by one the cats began to venture from their retreats as the afternoon progressed. Almost as if they thought that exploration is best accomplished on an empty bladder, most of them made their first port of call one of the brand-new tubs of sawdust, placed in every corner of the building—in fact, I got the impression that bursting bladders as much as curiosity finally forced them out of hiding. Among the adults, even cats who had formerly played and slept together took pains to avoid each other on these first tentative explorations; but on the narrow aerial walkways and at blind corners, encounters were unavoidable, and as a result the silence of the first couple of hours became increasingly punctuated by the howls and growls of scattered confrontations. These did not amount to much on the first day, the participants seemingly unready to commit themselves wholeheartedly to battle in the still unfamiliar surroundings.

The atmosphere had changed considerably by the next day. When I entered the building early that morning, several juveniles, including Marshmallow, Kenbo, Mambo and Ichibu, came bounding through from the far shed. Those possessing tails to speak of—on this occasion, only Kenbo and Ichibu —held them high in greeting (most Japanese cats sport stubs of various sizes and shapes, the longer ones often bent over at their tips like fishhooks, or twisted like corkscrews). However, they seemed to have better things to do than cling to my feet and were soon cavorting through the building in playful pursuit of each other. They surprised me with the confidence that they were already displaying as they leaped

The howls and growls of scattered confrontations.

up to and pounded along the aerial walkways. They had clearly wasted no time in getting to know the geography of their new dwelling, and their obvious appreciation of the space and scope for adventure that it offered was for me a gratifying sight: my labors had not been in vain.

Not surprisingly, the adults were not yet ready to join in the fun. One or two of the more timid cats, such as the female Chibi and the young tom Kenjiro, were still glued to the somewhat uncomfortable perches to which they had retreated the previous evening. I suspected that they had not budged all night. However, the majority of the adults seemed to have decided that the cattery presented no dangers in itself, and were moving around much more freely than on the first day. The females were generally peaceful, but from early morning we began to see real fireworks exploding between the large toms, Mo, Waru, Ee, Ganko and Sabu, while younger toms, such as Nibu, Panda and Sha, also became embroiled in arguments. Although the latter had no fighting experience whatsoever, they, too, seemed to realize that the time for reckoning had come. For the next three days there seemed to be hardly a moment when at least two toms were not facing up to each other. At night, even from behind the closed windows of the main house, over a hundred yards away, we could hear the howls emanating from the cattery. The toms fought mostly among themselves, but occasionally females were also the inadvertent targets of their aggression. Under natural circumstances, intersexual hostility is as rare among cats as it is among dogs, and the frequency with which it occurred in the cattery during that first chaotic period underlined the pervading tension and fear.

Perhaps the most dramatic incident that I witnessed on that second day was one such male/female confrontation, made all the more electrifying by the fact that the female involved, Aya, was the mother of her attacker, Sabu. Aya suddenly came charging through from the far shed with Sabu hot on her heels. Just in front of me, at the base of an old easy chair, she rolled desperately on her back to face Sabu, who leaped onto the arm of the chair to press his attack from above. Aya hissed and growled at him from below, and in her terror let fly a stream of urine. Sabu, staring menacingly down at her, seemed totally unaware of Pe's kittens, who cowered in the corner of the chair petrified with fear, their eyeballs almost popping out of their sockets. Chiyoko had moved these kittens into the cattery, together with their mother, Pe, a couple of weeks earlier, to give them time to adapt before the rest of the cats were transferred, but even with this period of acclimatization they were unprepared for such shocking happenings. Luckily for Sabu, and unluckily for Aya, Pe was elsewhere at the time; otherwise Sabu would have found himself on the receiving end of a barrage of claws.

As it was, Sabu continued to threaten Aya for about half a minute until, seeing no advantage in pursuing his attack, he turned, jumped off the chair and walked stiffly away, the hairs on his spine still bristling. Aya watched him disappear before she herself stood up and crept off to a quiet corner. Long before she felt safe enough to lick herself thoroughly, the kittens had started to play, almost as if to rid themselves of the tension created by the confrontation. The speed of this switch, from undisguised terror to abandoned romping, was remarkable. Over the following years, kittens were to impress me time and time again with such displays of emotional athleticism.

I, too, breathed a sigh of relief when Sabu eventually relented. During the next four years, but in particular during those first few days, my sensibilities were to be tested on countless occasions by such encounters. The whole idea behind the cattery was that the cats would be free to do exactly as they wanted, with myself in the role of detached observer. However, I had also taken on the project with the aim of improving the lot of the cats; I was now their keeper, responsible for their health and happiness. At times a conflict of these interests would be inevitable, but I realized from the start that emotion and partiality would serve neither my interests nor those of the cats. It was not always easy to remain impartial in some of the bitter, one-sided confrontations I witnessed, but I found in my cameras an unexpected ally. Had I not had my eye pressed to the viewfinder when Sabu threatened Aya, I suspect that I would have been sorely tempted to put the boot into Sabu on behalf of poor Aya and the

Sabu terrorizing his own mother, Aya. Pe's kittens are terrified onlookers.

terrified kittens. While I was acutely aware of the torment that these cats were suffering, the camera helped me to objectify the drama, and to realize that Sabu was as much a victim of circumstances (of my creation) as were the others.

While in actual confrontations I maintained a strict policy of nonintervention, treatment of injuries was a different matter. We could leave the cats to deal with light wounds to any parts of their bodies that they could reach with their own tongues, but we had to be very careful with the frequent head and neck wounds, especially during that first period: we could not yet trust the cats to lick each other's injuries and, in the new surroundings, full of new germs, unattended bites and scratches quickly turned septic. After feeding the cats each morning, Chiyoko and I checked them all for fresh battle scars, and tended a constant stream of torn ears and half-closed eyes, much to the outrage of the cats concerned. Until we became adept at immobilizing a furiously struggling tomcat, we often suffered as much as our patients; the tooth and claw marks that decorated our hands bore testament to the occasions when one of them got the better of us. Even now, I would prefer to deal with a protesting dog any day of the week, and I'm sure that many vets would concur with this sentiment.

At the time it seemed as though the blood and thunder would continue forever, but after a few days things began to quiet down. The big toms all looked weary and frayed around the edges. They fought less frequently and more selectively, and without the vigor that they had displayed at first. Brief, bitter exchanges of claws and fangs gave way to drawn-out cursing matches, the participants trying to browbeat each other into submission without recourse to actual physical contact. Increasingly I observed even the strongest toms deliberately avoiding situations of potential conflict with weaker opponents. Even at the height of hostilities, I had noticed that the toms who had formerly been living together peacefully did not fight with each other. I did not witness a single confrontation between Sabu, Nibu and Panda, or between Ee and Wanchan. Wanchan had been transferred to the upstairs landing as a kitten and had got on well with Ee, even after developing into a large, powerful young adult. Although he had had no fighting experience, on transfer to the cattery he quickly displayed a taste for aggression and actively challenged even the likes of Mo and Waru; but never a cross word was uttered between himself and Ee. In fact, within two weeks these two were bedding down together in the old trailer in the far shed, with Waru's sister Kurokasan, a black cat with lovely green eyes, often joining them as she had when the trio occupied the upstairs landing.

I also regularly found Mo curled up with his long-standing girlfriends, the sisters Uko and Aya, or with his own sister May. Sha frequently occupied the sofa with his sister Pe, and at such times attended to her kittens with much the same diligence that Pe herself displayed. Perhaps because it was the most comfortable spot in the building, this sofa soon became the focal point of life in the cattery, at times being occupied by more than twenty cats. During the first two or three days, very few cats had slept together, but by the end of the second week, only a small minority were passing the nights alone. Although they were almost certainly gathering for warmth as much as for company, daily observation revealed that, while various rivalries were developing among the tomcats, old friendships were being restored and new ones created.

Mealtimes were also far more relaxed. The cats could help themselves at any time to dry cat food placed in tins located throughout the sheds, but the high point of every day was their meal of boiled fish and rice, which they attacked with far more gusto than they did the cat food. The Kingdom being situated near a small but busy fishing port, Kiritappu (which translates as "Foggy Town" and was a very apt name), we were blessed with a regular supply of cheap fish from the processing plant situated there. This fish consisted mostly of forty-pound blocks of cod scraps plus, depending on the season, mackerel, mackerel pike and sprats of a grade considered too small or pockmarked to grace the average Japanese dining table. We took care to use the latter varieties sparingly, only to add a bit of "oomph" to the rather bland staple of cod, since there is a danger of cats developing an often fatal condition, known as steatitis or yellow fat disease, when fed large amounts of such fish, which are very rich in fats.

As the cats were already accustomed to

The sofa soon became the focal point of life in the cattery.

eating two or three to a dish, they showed no hesitation in continuing to do so in the cattery, once they had overcome their wariness of the place and each other. At first we had to feed the more timid cats separately, leaving bowls of food in front of their various retreats while the majority of cats ate below. After a couple of weeks, however, even those cats had gained enough confidence to join the others at mealtimes. And, whereas in the first few days the adults, in particular, had seemed in no mood to lick each other down after meals, most of them now gathered in groups of two or three, males and females alike, for communal grooming. The cats' appetites also seemed to have grown in direct proportion to the increasingly relaxed atmosphere and the much greater amount of exercise that they were getting. They had been used to only one fish meal a day, but by the end of the second week they were dispatching this with such voracity that I decided to double their fish ration and feed them both morn-

The gratifying sight of the early morning sunbathing session along the east-facing window shelf.

ing and evening. Even then, the tins of dry cat food were often empty by morning. I had weighed each of the cats on transfer and, on weighing them again after a month, found that the adults had put on an average of a pound, the juveniles even more. Despite the trauma of the move and the continuing tension and conflicts, almost all of the cats looked to be in much better shape to face the winter than when they had first set foot in the cattery.

No less gratifying than the way the cats polished off their fish each morning and evening with such obvious relish was the sight of them all sunbathing every morning on the shelf that I had fitted along the length of the east-facing window in the far shed. Cats are among the most fervent of sun-worshipers, and ours were no exception.

They had wasted little time in discovering this site, and would gather there as the sun rose and bask until we brought their breakfast. While the toms tended to fight most at this time of day, for some reason very few quarrels ever erupted on that shelf; it seemed almost as if the cats had an unspoken rule about keeping the peace there at that hour.

Toward noon each day, an almost funereal silence crept over the building. The product of thirty-odd cats taking their long midday siesta, this silence was a far cry from the one that had hung ominously over the place on the first day. Even though the power struggles between various toms continued, something approaching peace was beginning to penetrate the atmosphere. As an observer, I could at times have hoped for more action, but as keeper, I had good reason to be pleased with the way things were turning out.

3

THE RISE AND FALL OF A TYRANT

S*eptember 18, 1978, 7:15 p.m.* Waru up to his tricks again. First he chased Sha off one of the sawdust tubs in the near shed, just as Sha was preparing to defecate, and pursued him as far as the shed entrance. Sha scrambled up the chicken wire to the platform above the door, significantly his favorite lookout and retreat. Waru chose not to follow him up, instead strolling off toward the passage to the far shed. On his way, his eye caught Kenjiro, who, on seeing Waru approach, ducked under the low shelf of the table by the east window. Waru was clearly in the mood for more mischief and tried to get at Kenjiro from above and behind, but Kenjiro, while cowering, flattening his ears and growling, countered by changing his position under the shelf to keep his rear end from Waru. Waru padded around and around the table, crouching as he did so to get on the same level as Kenjiro, and then he suddenly dived into the attack. Kenjiro dashed out from the other side of the table and fled away, panic-stricken, with Waru in hot pursuit. There was a brief scuffle by the old massage chair before Kenjiro escaped under some boards. Waru thumped on top of the boards, sending Kenjiro skittering back to the location under the window table. Waru let him go this time, spraying energetically before disappearing through the passage into the far shed. During the whole contretemps he had not uttered a single sound. I got the impression he had enjoyed himself immensely. Not so Kenjiro, who had hissed and growled and screamed, but was now licking himself down while keeping an eye on the passage in case Waru reappeared. He had a cut on his right ear but otherwise seemed none the worse for the ordeal. A couple of cats stopped to sniff around the table and at Kenjiro's rear, but they were females and left him alone.

Waru in hot pursuit of a panic-stricken Kenjiro.

September 21, 1978, 7:30 a.m. Ganko was strolling around at the end of the far shed, as was his habit, when Waru spotted him. He approached slowly and silently, unnoticed by Ganko, who disappeared under the jeep. Ganko had climbed up onto the rear axle and was sniffing the undersurface of the chassis when Waru crept up behind him. Waru momentarily stretched his nose toward Ganko's rear. He was half squatting, and I could see his tail twitching slightly. He clearly intended to give Ganko a surprise nip on his rump but he deliberated a little too long. Ganko sensed Waru's presence and whipped around, uttering a high-pitched growl as he dropped off the rear axle. Waru jumped backward in surprise, banging his head on the muffler, which gave him another shock. This double bungle seemed to have shaken him and made him lose his nerve, for he slowly backed away from Ganko, who was growling and threatening him with more authority. Waru rounded the outside of the jeep and tried to approach Ganko once more, but after a further brief exchange of curses, Waru walked away slowly and stiffly, his expression still threatening even as he retreated, until he had put a safe distance between himself and Ganko.

These are just two of many incidents recorded in my observations of that early period, in which Waru featured prominently. The fact that his name appears on virtually every page of the notes I took will give some idea of how conspicuous he made himself. He seemed to be everywhere, the busiest bundle of "bovver" in the cattery, and for this reason has earned himself a special slot in my memory.

Waru was born in the summer of 1976, the product of Mo's courtship with Tobo. He grew up in the television room, but was moved to the living room when trouble began to flare up between him and Mo the following summer. It was around this time that he came to be called Waru, which means, very simply, "evil," since he seemed to create trouble wherever he was put. Mo had got the better of him, but in the living room the old tom Brapants (who died in the winter of 1977) and, later, Ee upstairs had had such problems with him that Chiyoko transferred him to her room, leaving him in charge of the juveniles Sha and Pe. By the next summer, as preparations for the cattery were under way, Pe was pregnant with his kittens while Sha was increasingly the target of his spite. For Sha, the opening of the cattery meant salvation, as there he became only one among many other toms whose existence Waru found objectionable. During the general hostilities of the first days, Waru had been no more conspicuous than Mo, Ee, Sabu or Wanchan, but within two weeks he seemed to be the only tom who had not tired of doing battle, and who regularly went out of his way to create trouble.

Over the next month or so I observed the first and, as it turned out, the last signs of a rudimentary territoriality among some of the tomcats. For example, Ee, as mentioned earlier, had developed an attachment for the trailer in the far shed, and Ganko for the area at the end of that shed. Although he often joined Ee in the trailer, Wanchan was often seen atop a structure in the near shed which I referred to as the "tower," consisting of four upright logs joined in a square by various platforms and ladders. Sha also had a favorite spot, the platform above the entrance to the near shed. I call these territories rudimentary because they were only a passing phenomenon, and because the cattery was not nearly large enough to allow the establishment of proper territories. Even among free-ranging cats, territories are by no means hard-and-fast entities, tending rather to expand and shrink like vast amoebae, according to season, time of day, the activities of neighboring cats and so on. Tomcats establish far larger territories than females, and it is probably for this reason that their territo-

Ee drives Waru from his territory, the trailer.

ries show a great deal of overlap. Trespass on another cat's territory is not necessarily an invitation to attack, often being permitted by cats on good terms with each other. Moreover, a very strong tomcat can often come and go as he pleases within the territory of a weak opponent. However, even a weak tomcat may display far more willingness to face up to an opponent within his own territory, particularly in the center of his range, with this resolve decreasing as distance from the center increases.

This phenomenon, of strength related to location, has been given the term "relative hierarchy" by Hediger, as opposed to the "absolute hierarchy" of fixed ranks often found within groups of social animals and exemplified by the pecking order among flocks of hens. The toms I have mentioned showed unmistakable signs of relative hierarchy during that period. For example, Ganko usually avoided Waru like the plague, but in the instance described from my notes, he faced up to Waru with admirable resolve under the jeep, which was significantly located in the area at the end of the far shed where he habitually patrolled. Also significant was Sha's escape, on being pursued by Waru, to the platform above the shed entrance. Despite his magnificent physique,

Mutsugoro's Animal Kingdom from the air.

The sad facts of life in the Kingdom—dogs lived outside, cats inside.

Sha and Panda facing up to each other during the traumatic first few days in the cattery.

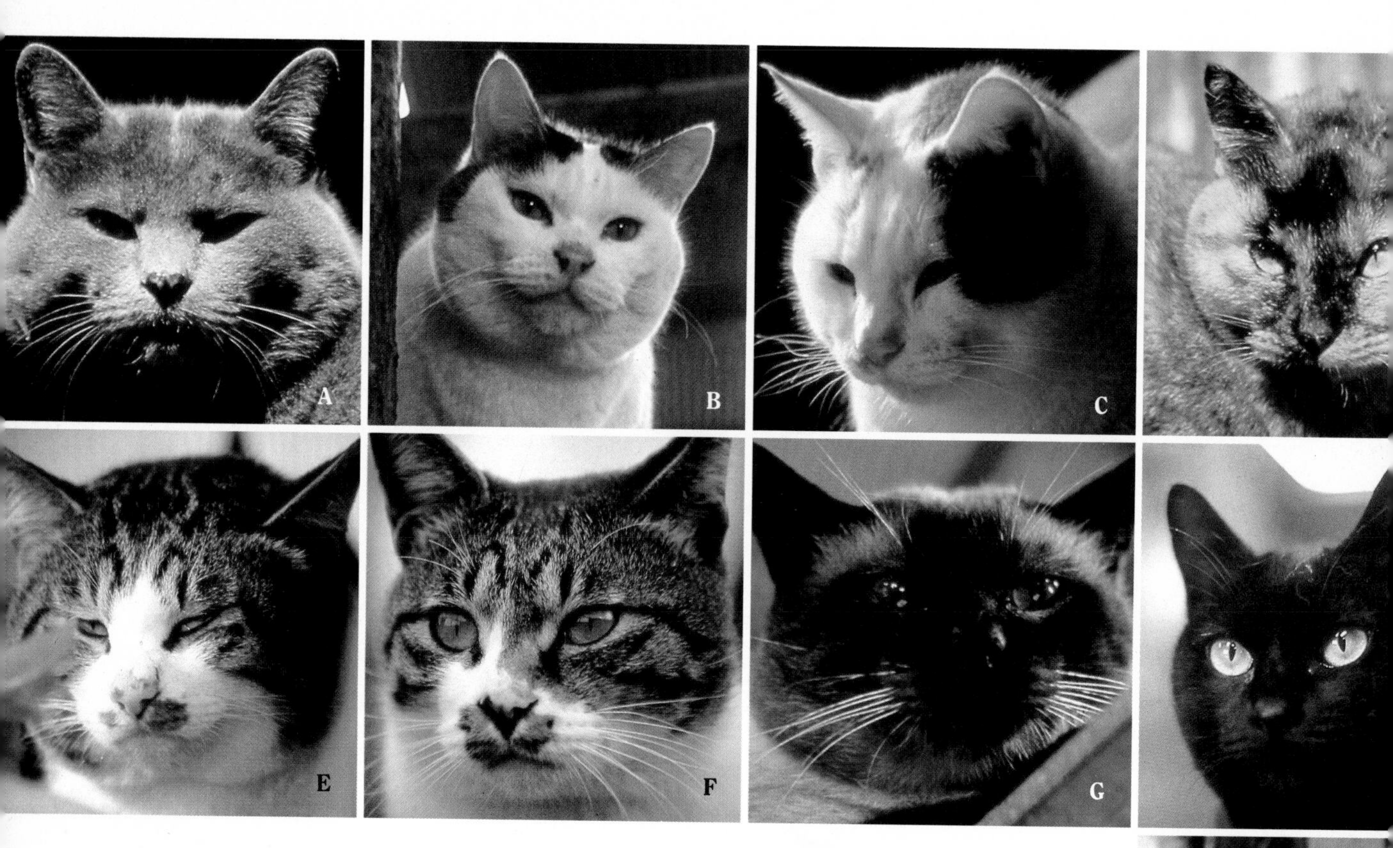

A MO (b. 1973), *son of Ma, the founding father of the Kingdom cat population. Now the oldest cat in the cattery, but hardly shows his age and is still number one tom. No despot, he is the perfect gentleman toward females. More than 70 percent of the Kingdom cats probably contain his blood.* **B, C** AYA *and* UKO (b. 1974), *sisters who came to the Kingdom from a nearby farm at the age of six months. Became members of Mo's harem and have borne many of his kittens. Alike in many ways, they are very gentle and tend to be bullied by other females. Neither can stand toms dueling in their vicinity and sometimes butt in to break up a fight.* **D** PE (1976–84), *found as a kitten by Chiyoko with three siblings stranded on a rubbish dump. Reared with an Ainu hound, she stood no nonsense from dogs. Strong-willed, intelligent and playful, she was very popular with the toms and an excellent mother.* **E** NIBU (b. 1978), *son of Mo and Aya. In the main house, apprenticed to his older brother Sabu, he got his first taste of love and has never looked back. The ultimate sexual technician but, fortunately for his fellow toms, he soon bores of the same girlfriend.* **F** SABU (1977–80), *son of Mo and Aya. Almost unbearably affectionate as a young cat, but after moving to the cattery had no time for people. Busiest sprayer I have ever known. Sadly, got out of the cattery one day and was killed by the dogs.* **G** WARU (b. 1976), *son of Mo and Tobo. As a youngster he relished fighting and was good at it—the only tom to get the better of his father, Mo, during the first two months in the cattery. Got his name, meaning "evil," by causing trouble wherever he was put.* **H** KUROKASAN (b. 1976), *daughter of Mo and Tobo, sister of Waru. Her name, given to her after the birth of her first kittens, means "black mother" and suits her well, her heart being almost as black as Waru's at times. Bullies other females when the mood takes her. Lovely green eyes are her strong point.* **I** KENBO (b. 1978), *son of Mo and Uko. Usually a rather weak-natured and inconspicuous character, but a hunter whose bravery is matched only by that of Marshmallow. Famous for catching the first sparrow to get into the cattery, a feat he has never forgotten.*

J MARSHMALLOW (b. 1978), *son of Mo and Uko. Small in stature, he is a matchless rat catcher and the only one who will tackle without hesitation the monster rats sometimes caught in the traps we set. He is also the only one crazy enough about prey to try to steal some other cat's rat.* **K** WANCHAN (b. 1977), *son of Mo and Uko. Big and strapping, with an aristocratic air, but the emergence of peculiar sexual habits have made him increasingly the laughing stock of the cattery. Like a younger brother to Ee.* **L** PANDA (b. 1978), *son of Mo and Aya. Was discovered at birth to have a slightly deformed tongue which prevented him from suckling, so had to be hand-reared. Perhaps as a result, he will eat anything and everything—except pickled radish. His black, gray and white fur is strange, but unlike most tortoiseshell toms he is very fertile.* **M** SHA (b. 1976), *brother to Pe and hand-reared with her and the Ainu hound. Very easygoing, he does nothing in a hurry, which may be why he is the heaviest cat in the place. Has formidable fangs which shine elegantly against his black face when he bares them in a duel, but his voice lets him down—he is still a soprano. Hates Panda.* **N** EE (b. 1973). *"Donated" as a kitten by the local branch of the Farmers' Union, he was not at first a great fighter, but after being nipped by Waru while mating, he began to show he had some muscle and resolve beneath his podge and gentle nature.* **O** TOBO (1973–81). *Picked up after a blizzard with a nasty sore on her upper lip, which lost her her looks, but not her popularity with the toms, who go by smell, not appearance. Until she died of cancer she reared a host of kittens, both her own and those of other cats. Loved and remembered by all who knew her.* **P** GANKO (b. 1976), *infamous for bringing fleas into the Kingdom. Whether due to shyness or to being a born gentleman, he is a very courteous lover. He has a protruding belly button, but it does not seem to bother him.* **Q** UEMI (b. 1976), *daughter of Kurokasan and Ee. Born with a defect in her eye musculature which makes her eyes turn permanently upward. Lost a lot of weight in the first cattery spring and killed two strange kittens, but regained her health after a period of isolation and became a perfect mother. Rather a bully, like her mother.* **R** HIGE (b. 1978), *son of Mo and Kurokasan. Earns his place on this page simply because of his funny face. His name means "beard" or "moustache." His resemblance to Groucho Marx made it difficult to think of him as a cat. Small and rather pugnacious, he regularly gets beaten up by bigger and older toms and takes his revenge on those younger and weaker than himself.*

With the huge increase in space, the cats were getting much more exercise and working up appetites for their next meal. They would congregate expectantly as mealtime approached.

"Bovver Boy" Waru on the lookout for trouble.

As life in the cattery settled down, old companionships were renewed and postprandial mutual grooming sessions were a reassuring sight.

Sha was still young and no match for the likes of Mo or Waru, but since the inception of the cattery he had developed a distinct dislike for Panda and fought with him frequently. I witnessed two occasions during that period when Sha, from his location on the sofa, saw Panda advancing along one of the walkways leading to "his" platform and purposely roused himself to scramble up onto the platform to send the weaker Panda hurtling back in the direction he had come. If Waru had followed Sha up to the platform, I strongly suspect that Sha would have made a stand there; the reason that Waru had not pursued Sha any further was probably that he was wary of engaging any opponent up high, and particularly on the narrow aerial walkways. Up until then I had seen him fall three times in the middle of scuffles which took place at such locations. For some reason the rogue was not blessed with the most perfect coordination and was prone to accidents. Although the situation was different, bumping his head on the jeep muffler when he confronted Ganko seemed somehow very typical!

Ee and Wanchan also showed far more inclination to fight when on their respective turfs. At that time Ee and Waru were mortal enemies and fought frequently, usually at Waru's instigation. Most of the tomcats could wander through the trailer without drawing a response from Ee, but despite his inferiority to Waru, he would not hesitate to confront the latter on or near the trailer. Ee's insistence on defending his turf appeared to peeve Waru; he passed through or near the location many times a day, always on the lookout for Ee, seemingly with the express purpose of upsetting him; and as a result I saw a number of dramatic exchanges between the two of them.

In comparison with these "territorial" toms, neither Mo nor Waru showed attachment to any particular location within the cattery. Both were ready to face any opponent anywhere, but of the two, Waru showed far more ambition. He gave the impression of being driven by dreams of domination. At two years of age, he was in his prime, a powerful and handsome brute of a cat. Whatever he lacked in grace he made up for in sheer determination and cunning. He seemed to delight in taking opponents by surprise, creeping up to deliver a painful nip to tail or rump when his unfortunate target was sleeping peacefully or squatting in a sawdust tub, face to the wall. One day he caught Mo like this at a bowl of cat food. Mo let out such a scream that every cat in the place fled in all directions up to the rafters. Within a matter of seconds the floor was completely cleared of cats—with the exception of Waru, who remained standing in the same spot, looking somewhat bewildered by the commotion.

He also showed no hesitation about butting into other cats' quarrels. On one occasion Mo tussled with Sabu in an old saddle hung from the ceiling, as a result of which Sabu slipped and was hanging by his foreclaws to the bedding inside the saddle, his rump suspended about two feet above the floor. Who should be passing below just at that moment but our illustrious bovver boy?

Waru creeps up on an unsuspecting Wanchan, and nips his rump.

It was a golden opportunity for Waru, who, calm as you please, stopped in his tracks, stood up on his hind feet and embedded his fangs in Sabu's rump. Poor Sabu screamed, fell to the ground, scuffled briefly with Waru and then fled. Waru watched him disappear and then proceeded on his way as if nothing had happened.

In view of such tactics, it was hardly surprising that most of the cats came to connect Waru with trouble. Even the females, whom he never attacked, watched his movements nervously and occasionally went for him with their claws. In the cattery, which was in general becoming more peaceful with every passing day, Waru was the one big source of action and trouble. By the end of September there was only one cat, Mo, who seemed willing to face up to Waru anywhere, and it was toward Mo that Waru was increasingly directing his spite. Where these two cats are concerned, I have to admit to some partiality. I was much impressed by Waru's strength, energy and determination, and amused by his low cunning and occasional blunders. I was also grateful for the interest that his presence added to life in the building. Mo was, however, the one cat whom I really did not want to see capitulating to Waru's tyranny. At six years, he was still far from old, but he was nevertheless the oldest tom in the Kingdom. Although he had been undisputed master of the television room, I had been concerned as to how he would fare among a host of younger rivals in the cattery. After all, the older an animal is, the more difficulty it usually has in adapting to a new environment. As things turned out, however, Mo soon proved to be more than a match for any young upstart, including Waru. And although he trounced his opponents with consummate confidence and authority, he possessed none of Waru's malevolence and was not the despot that he had at times appeared to be in the television room.

As the cattery entered its second month, Waru was, with his persistent challenges, giving Mo pause, but I was sure that Mo would prevail. Then, on the morning of Oc-

By the end of September, only Mo seemed able to stand up to Waru.

Mo cowers before a victorious Waru.

tober 17, I witnessed for the first time the sad sight of Mo in flight, Waru at his heels. Mo turned to defend himself at the entrance to the overhead passage joining the two sheds, but his posture was no longer even that of an equal. He cowered and growled—no longer a growl of authority but of fear—while Waru glared down at him. In contrast to Mo's appearance of a thoroughly defeated cat, Waru's was that of a victor brimming with confidence. He uttered not a sound, and his ears were pricked upward and backward, not at all flattened as Mo's were. Although there were no signs that the two had clashed physically, and although he left Mo alone after standing over him for about half a minute, Waru had made it clear who was now on top. He had finally succeeded in overcoming his last and strongest opponent, but I was in no mood to congratulate him.

The axiom that what goes up will sooner or later come down again, thankfully applies as much to despots as it does to most other things on this earth. Waru's fall was by no means sudden, and it is difficult to say exactly when it started, but I like to think that the extraordinary events of the night of December 4, 1978, were critical in starting the process. That night was remarkable for the fact that a young female called Wan Two, a sister of Wanchan, had unexpectedly come into heat. Cats living in the northern hemisphere, far from the equator, are not expected to come into heat until at least a couple of weeks after the winter solstice, since it is the lengthening of the days which is thought to trigger the hormonal mechanisms that bring on estrus. I suppose that Wan Two was the exception that proved the rule. Throughout the day she had been pursued by a crowd of toms consisting of Mo, Ee, Wanchan, Sabu, Nibu and Panda, but had spurned their advances until the evening. As this was the first genuine sexual activity to occur in the cattery, both Chiyoko and I stayed up late to watch the proceedings, which so engrossed us that we hardly noticed the passage of time. The two things that most surprised us were, first, the lack of argument among Wan Two's suitors, and, second, Waru's conspicuous absence from their ranks. In our ignorance we had expected the strongest tom to monopolize any receptive female, especially a tom as intolerant as Waru.

The tom in question finally showed up close to midnight, by which time the romantic action was taking place by the table tennis table in the far shed. Ee was on top of Wan Two when Waru appeared, with the rest of the toms gathered in a circle around them. The spectating toms seemed to be too engrossed in the performance to notice Waru as he stole up behind Ee and Wan Two. His expression was one merely of mild interest as he sniffed Ee's hindquarters. But then, with no warning whatsoever, he performed his favorite trick, suddenly latching his jaws into a fold of Ee's rump, adding his claws for good measure. No act could have demonstrated more aptly Waru's appetite for mischief and his total contempt for gentlemanly conduct; almost as if they also thought so, those toms who had not fled immediately upon the eruption of violence turned in unison to face the enemy. Although I risk accusations of anthropomorphism, I detected in their expressions and tones of voice the kind of righteous indignation that could only spring from those who have suffered the gravest of insults and injuries. At any rate, Ee and Nibu, who threatened Waru with the most resolve, displayed a degree of courage which had been totally lacking from their recent confrontations with the tyrant. Young Ni-

Waru goes too far. Creeping up on an otherwise occupied Ee . . . he commits his dastardly crime . . . Ee and Nibu confront the villain . . . and Waru retreats.

bu's reaction in particular surprised us: although he had never been a match for Waru, he forced the culprit to retreat step by step with a constant barrage of menacing growls. Waru clearly wanted to have nothing more to do with the situation, but it was some time before Nibu tempered his threats sufficiently for Waru to turn his back. He seemed to have been totally nonplussed by the reaction of the toms, most of whom, including Ee, had now recovered their composure and were once more hot on Wan Two's trail.

Waru was still a formidable opponent for some time after this incident, but it seemed to me that the other toms learned that night that he was not invincible. While they never went out of their way to bandy words with

Mo once more became "top cat," and never let Waru forget it.

him, they faced up to him with increasing confidence when challenged. However, it was almost certainly Mo who added the most momentum to Waru's fall. One mid-January morning Chiyoko burst into my room to announce that she had just seen Mo and Waru have a terrific tussle.

"It's a long time since I've seen such a fierce fight. I'm telling you, fur was flying in all directions," she related excitedly. "In the end Waru was on the floor, all four legs in the air to hold Mo off. The poor devil looked completely KO'd."

After that, hardly a day passed without Mo letting Waru know who was number one, although physical exchanges between the two were infrequent. Waru was always on the lookout for Mo, creeping behind furniture and crouching quietly whenever Mo passed close to him, and it was this nervousness about bumping into Mo which hastened his retreat to the fringes of life in the cattery. While he never became a pariah, and continued to be an object of fear among the other toms, he was by and large ignored. He seemed to be afraid to join the groups of toms that always accompanied females during the first two or three days of the latter's heats, contenting himself with females in the last stages of estrus who were no longer in much demand. He no longer sprayed with the vigor that he had shown in his heyday, when my observations revealed a spraying frequency approached only by Sabu. At times he seemed so apathetic that I thought there was something physically wrong with him, but investigation revealed nothing but fleas, which were endemic in the cattery despite regular powderings.

I had not been sad to see Mo triumph over Waru once again, and felt that the tyrant more than deserved whatever punishment and cold-shouldering he received from the other cats. At the same time, however, I couldn't help feeling sorry for him. He was, after all, a victim of the circumstances I had imposed; he was clearly ill-adapted to life in the cattery, with the degree of social tolerance that it required. In a more natural free-range situation, I suspect that he would have thrived, if his first three months in the cattery were anything to go by. As it was, he was increasingly a shadow of that former self, and ceased to figure prominently in my records. I'll finish his story with an extract from my notes for June 20, 1980:

Waru came up to me this morning as I was cleaning the sawdust tubs. He rubbed around my legs, purring and begging for caresses. He is very friendly these days, seems almost contented. In the old days he had no time for such attention, in fact he positively avoided us, as he objected strongly to the treatment of his frequent scars. While I am flattered by his friendliness, I can't help feeling nostalgia for the villainous young tom that was Waru in the old days, and almost wish he would oblige me once again with a sneak attack on the backside of an unsuspecting rival. He hardly deserves the name "Waru" nowadays.

Motherhood and kittenhood—both exhausting occupations!

4

ORPHANS

"Let's operate, then, and save ourselves any more worry," said Iwase-san, the Kingdom's resident vet, as we watched over Chibi in the delivery room. It was a cold night in early March 1979. Chibi, a beautiful orange tabby, was the second cat due to give birth in the cattery since we had transferred the cats there. Her pregnancy had proceeded normally, surprising us in view of the fact that she had miscarried twice in the previous year. The trouble was that she was now, according to my calculations, at least three days overdue, an estimate which the tautness of her belly, distension of her teats, and her increasing restlessness and appearance of discomfort suggested was not far off the mark. I had been waiting for two days for her labor to start, and was getting very anxious. Iwase-san suspected that a fetus, either dead or alive, might be jammed in the birth canal. If so, any further delay posed serious risks to the lives of both Chibi and her kittens, and when an injection of oxytocin failed to do the trick, Iwase-san decided on performing a cesarean delivery as the safest solution.

Preparations began immediately, and within half an hour Chibi was laid out anesthetized on the living room table, with six of us gathered around to lend Iwase-san a hand. A few minutes later, he was extracting the first of the kittens from Chibi's womb, with four others following within minutes of each other. As Iwase-san had feared, they showed little sign of life. One person to one kitten, we all started to work furiously on them, first sucking their noses and mouths clean of amniotic fluids, and then giving them the kiss of life and chest massage while rubbing them dry in front of heaters. The kitten that I was tending and one other soon began to respond, their breathing albeit irregular at first. The other three, however, appeared to be goners, showing not a spark of life even after ten minutes of treatment.

"Goners? Nonsense! It's still too early to judge, so stop worrying and keep at it!" Iwase-san scolded any pessimists while he attended to a very weak Chibi. And sure enough, miraculous though it appeared to us at the time, about five minutes later first Hiroko and then Chiyoko joyfully reported that their kittens were beginning to breathe of their own accord. Finally, after working silently and intently on his kitten for twenty-

Chibi's five kittens, not exactly lively, but nevertheless alive.

five minutes, determined as always not to fail where his seniors had succeeded, the horse-crazy seventeen-year-old high school dropout, Kazuhiro, known to all of us for some reason as Otochan ("Daddy"), laid his kitten alongside its four siblings in front of the heater. All five, though still showing little motion, were breathing regularly, and when Iwase-san had done all he could for Chibi, we placed her in one box, the kittens in another, and Chiyoko took them up to her room.

Tragedy struck the next morning. Chiyoko, awakened by the cries of the five hungry kittens at about six o'clock, found that Chibi was dead. In her distressed condition, the operation had been too much for her. Shocked and saddened as we were, there was little time for somber reflection with five hungry orphans suddenly on our hands. They received their first feed soon after Chibi's death. The one good thing about not having had a chance to suckle on their mother's teats was that they took to formula milk from a feeding bottle right from the start. Many of us in the Kingdom knew from experience the difficulties that are often encountered in getting a baby animal to accept a feeding bottle once it has learned the sensation of its mother's teats and the taste of her milk. The longer it has been with its mother, the greater these difficulties are, but even a baby which has only once fed successfully from its mother will often stubbornly refuse any alternative, despite its hunger. The next year, when Chiyoko took on the care of three kittens who had been rejected by their mother after the first day, she had to force-feed them for three days before they finally began to drink voluntarily from a feeding bottle.

While Chibi's kittens posed no such problems, the bad thing about having missed out on their mother's teats was that they would be lacking the antibodies contained in the colostrum, which protect newborns from a variety of diseases during the first few weeks of life. We thus had to watch the orphans' health very carefully. However, a more immediate problem was constipation, which we discovered invariably developed in orphans raised from birth on formula milk. The kittens readily urinated when their hindquarters were stimulated with the baby-oiled tips of our fingers, but strain though they did, they had increasing difficulty expelling feces, and we found it necessary to give them enemas once a day until they started to eat solid food.

When I volunteered for the cattery project, Chiyoko had agreed to assist me on one condition: that she would at some time get the chance to hand-rear some kittens from birth. She reminded me of this condition on the day of Chibi's death; while she had no objection to sharing the experience, she insisted that the kittens remain in her room. I soon discovered that, where maternal instincts were concerned, she had me beaten hands tied. Although I camped down in her room with the box of kittens at my head, she proved to be far more sensitive to their cries in the dead of night than I was. When I woke at all, it was frequently to find Chiyoko already tending the kittens, who, during their first week or

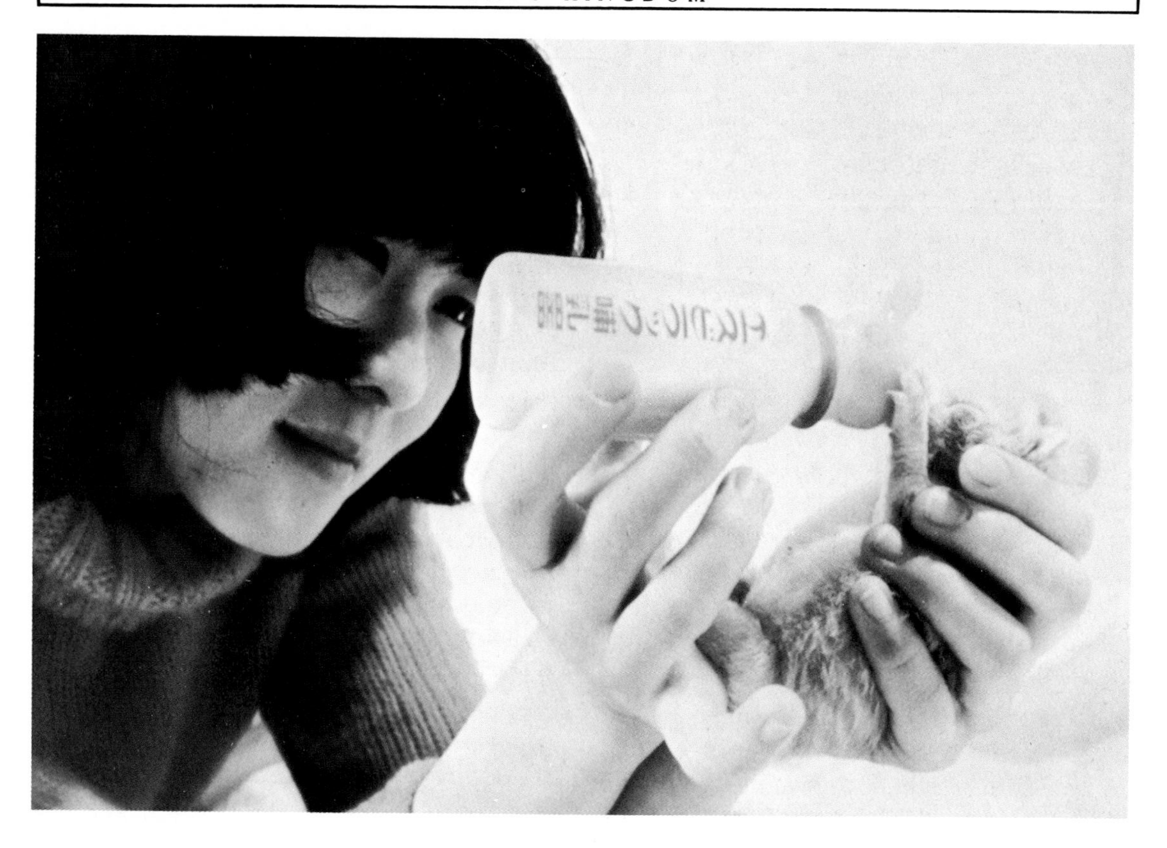

One could literally see and feel the kitten's belly filling out as it drank.

so, were fed every three or four hours both day and night.

However, when I was both present and awake, I shared the trials and joys of feline parenthood with Chiyoko. Where baby kittens are concerned, the joys to be found in feeding are considerable; cupping the tiny creature, still blind and deaf, in the palm of one's hand, one can literally see and feel its belly filling up as it drinks so industriously. After feeding, we would find ourselves staring down dotingly on those five rotund bellies pressed snugly to each other as their owners slipped back almost immediately into deep sleep—and who could blame us? We were, after all, their parents, as involved in their upbringing as any mother cat with her newborn kittens. We got as much of a kick out of caring for them and following their growth day by day as we do now with our baby daughter.

At two weeks, as soon as the kittens showed the first signs of play and had begun to walk shakily around inside their box, we opened a hole in it to a larger open-roofed box which would serve as dining room, playground and lavatory, with a small tub of sawdust in one corner. Despite the pleasure that we got out of bottle-feeding them, we wanted to get them onto solid food as soon as possible to put an end to their constipation. They rose magnificently to the challenge, stumbling out of their nest in response to our calls, and straight into bowls of fish gruel. Licking their paws clean, they soon began to appreciate the taste and bury their faces where their paws had trodden. Although their method of ingestion was at first more akin to sucking than eating, and got them so plastered that daily baths (our only answer to a mother cat's tongue!) were in order, within a week they showed a lot of improvement and were largely supporting themselves on this food at an age —about three and a half weeks—when many kittens undergoing a more natural upbringing still don't know of the existence of any sustenance other than their mother's milk. The gruel had the required effect of moving their bowels, and after we had scented the sawdust by stimulating them to urinate and defecate over it a few times, they began to toddle off to the tub of their own accord upon waking or after eating, to perform there.

By this time, they had long since received names, based on those of the people responsible for bringing them to life on the night that they were pulled out of Chibi's womb. The tortoiseshell female and one of the orange and white males were named respectively Toco and Hi, after Toshiaki Ishikawa and his wife, Hiroko. The other orange and white male had the dubious privilege of inheriting young Kazuhiro's nickname, Otochan. The remaining two, the orange male and tortoiseshell and white female, became respectively Jerry and Chi after ourselves. The five kittens came to be known collectively as Otochantachi ("The Otochans"), after Otochan, who was the biggest and invariably the first to try anything new.

Otochantachi rapidly outgrew their boxes and took over the whole room (which, at about eight square yards, was, admittedly, not expansive—a typical Japanese rabbit hutch), barring, for the time being, the built-in cupboard (or *oshiire*) where Chiyoko slept.* I would drop in in the evening to find Chiyoko lying at play with the kittens among a jolly sea of old clothes, toys, dolls, dried flowers, balls of wool, crumpled paper and anything else in the room that could be knocked down or pulled out onto the floor by Otochantachi, whose ability to create chaos and feel completely at home in it was no less impressive than that of any other kittens. Their games were mostly destructive, in result if not in intent, but one of their favorites, "lighthouse," was quite inventive. Any Westerner visiting an aver-

*These cupboards, known as *oshiire* (literally "push-in"), are thankfully expansive, being designed to have Japanese mattresses (futons) "pushed in" to them by day, and in general any other belongings that won't fit elsewhere into the average, very poky room. They are normally split into two levels, and Chiyoko, departing from tradition, saved herself the bother of pulling out her futons each night and pushing them back in the morning by spreading them out permanently on the two square yards of the upper level.

age Japanese household will be struck not only by the cavernous *oshiire* but also by the complexity (and ghastly design!) of many overhead lighting fixtures. In addition to a main wall switch, there is invariably a string hanging from the fixture itself, successive pulls of which will offer different strengths of lighting, including a night-light and an off-mode, enabling the average floor-dwelling Japanese to operate the lights without having to get up each time.

Chiyoko attached a little knitted teddy bear to the end of her light string, and the kittens soon discovered that, by leaping and grabbing at this, they could turn the lights on and off, an activity which could keep them entertained for up to an hour, but would drive me crazy after a few minutes. Chiyoko was far more tolerant of this and their other pranks, but even she began to reach the end of her tether when Otochantachi became able to scramble up into her cupboard bed and play tag around her at all hours of the night.

One day she announced that, since spring was upon us and the days were growing steadily warmer, she was going to familiarize the kittens with the outside world, and put them through such adventures during the day that they would give her some peace at night. This was more an excuse to indulge the kittens than anything else, since Chiyoko knew as well as I did that, even if the kittens exhausted themselves during the day, they would be up to their highjinks again by midnight. I applauded the idea anyway, since I was no less eager to indulge the kittens, and also wished to indulge myself in the pleasure of photographing and observing kittens in natural surroundings for a change. Otochantachi were ideal subjects, being just at the age when a real mother would begin to take them on hunting expeditions. Regarding us as parents, they would follow us readily, depending on rather than resenting our presence, as even the homeliest of adult cats is likely to do when out wandering.

Of their outdoor adventures, I think that the photographs speak largely for themselves. Needless to say, outings became the high points of each day for the kittens. They soon connected the wicker cat basket we bundled them into with adventure, and would jostle each other in their haste to get into it when in Chiyoko's room, but rounding them up again when it was time to return home was a very different matter.

We took them to woods, fields and pastures, and even once to the beach. The latter was, I must admit, hardly the kind of location that a cat would naturally choose, and Otochantachi were at first predictably terrified by the lack of cover, the immensity of the landscape and the roar of the waves. Once over the shock of suddenly finding themselves in such surroundings, however, they responded bravely to our calls of encouragement, and came trotting along the beach after us, even going so far as to wet their feet in the cold surf of the Pacific Ocean.

We actually did very little walking on these outings since, unlike dogs, cats are totally unsuited, both physically and mentally, to prolonged strenuous exercise. De-

Waiting to be taken out.

spite their speed and agility, they lack stamina and tire easily; and although they may be willing, as Otochantachi were even as adults, to follow human companions, they much prefer to explore at their own leisurely pace. This is, of course, only to be expected of silent and solitary stalkers of small prey.

It was thus far more satisfactory for the kittens if we left them to their own devices on these outings — and far more rewarding for ourselves. By getting down on all fours and following them through grass and undergrowth, stopping where they stopped and looking where they looked, our attention was drawn to a host of things we might otherwise have missed, and we gained a closer appreciation of the world as they might have perceived it — a world of close, obstructed horizons in which the senses of hearing, touch and smell were every bit as important as sight; a world populated with all sorts of tantalizing little life forms to watch, chase, capture, paw at, chew and, if the taste was agreeable, to eat.

Comfortably settled on the back of the Percheron mare, Hoshiko.

By autumn the kittens had familiarized themselves with tadpoles, frogs, stag beetles, butterflies, moths, dragonflies, crickets and shrews, and also with much larger animals such as the Kingdom's horses, sheep and pigs. These they approached of their own accord, diffident but driven by their curiosity. Sometimes they would be driven away, but more often their noses, extended in greeting, would be met by those of the equally curious strangers. We rarely forced these encounters on the kittens, but one winter's day, when they were already about nine months old, we could not resist lifting them onto the warm back of our Percheron mare, Hoshiko, on the way back from a walk in the snow-carpeted woods. Since we fully expected them to jump off immediately, you can imagine our surprise when instead they settled down in the thick winter fur of Hoshiko's ample back and proceeded to lick each other down, only dismounting in a hurry when Hoshiko, who didn't seem to mind the imposition, ambled off to a water trough.

Although we continued to take Otochantachi on outings, for their sakes and Chiyoko's we transferred them to the cattery when they had reached the age of three months. Probably as a result of the adventures they had been through up until then, they showed not the slightest fear at finding themselves among so many strange cats, and started to play almost immediately with the other kittens that they found there. From that day on, apart from their outings, they led the same lives and grew up in much the same way as their cattery contemporaries. When they began to test their fighting skills the next spring, Otochan, Jerry and Hi interestingly appeared to prefer crossing swords with each other rather than with other toms of the same age. I subsequently noticed the same tendency among other young toms who had grown up together as littermates.

The same did not go for Chi and Toco, however. As I shall describe later on, they grew up to prove themselves both exemplary mothers and sisters who reared their kittens together in two successive years. I should add that, when they gave birth, both Toco and Chi made us more than welcome. Indeed, Chi refused to settle down and nurse her kittens for the first couple of days unless either Chiyoko or I was with her. All five cats never lost their strong attachment to human beings and, from all I hear, perform to this day the role of welcoming party to any visitors to the cattery. They were always the first to run to greet visitors, delighting many but genuinely frightening a few out of their wits with their habit of leaping suddenly out of the blue onto shoulders. If allowed to, they would willingly remain perched there to accompany guests around the cattery and pose for photographs.

On reflection, Chibi's tragic death had brought unexpected benefits. We fulfilled our obligations to her by rearing her kittens successfully, and the knowledge gained in the process proved useful time and time again, both to the other members of the Kingdom and to the many outsiders who called for advice on the rearing of orphaned puppies, kittens, even fox and tanuki cubs which had come into their hands. But perhaps Chiyoko and I gained most from the experience—many cherished memories of our five foster kittens exploring the Hokkaido countryside, and an invaluable foretaste of the delights and tribulations that real parenthood held in store for us.

The garden provided a whole host of new experiences.

5

THE MATING GAME

In the spring of 1979 I made a couple of improvements to the cattery, which I hoped the cats would agree significantly raised the quality of their lives there, and which I know vastly improved the quality of photographs that I was able to take of them.

The first of these changes was the introduction of light, or rather much more of it, to the sheds, which I accomplished by carving sections out of the corrugated iron roofs and fitting transparent plastic sheets in their place. Up until then the sheds had been depressingly dark; to satisfy their sunbathing appetites the cats had had to make do with the small amount of sunlight which slanted in through the east-facing windows in early morning and the west-facing windows in late afternoon. At other hours of the day, the sheds were as dark on sunny days as on cloudy ones. The skylights, about thirty in all, made a world of difference: although still a roofed and walled environment, the building no longer felt like a cavern, and the cats were now able to spread themselves here and there on innumerable perches that caught the sun at various times of day.

As regards photography, any cameraman would, I think, appreciate the joy and freedom I felt at being able to discard heavy flash equipment, at least during the daylight hours. It was still several aperture stops darker than outside conditions, but nevertheless sufficient for nonflash shooting in most situations, provided I maintained a steady grip on the camera. Photographing the cats became at once far more enjoyable and far more challenging, and the results were infinitely more natural and pleasing to the eye. The one disadvantage was that the cats tended, particularly on sunny days, to snooze through more daylight hours than ever, as a result of which I amassed an overabundance of snaps of luxuriously sunbathing, sleeping animals.

The second improvement was appreciated by the cats as much as the first. On the death from old age of the three-legged wild boar, Bui, we wasted no time in commandeering his enclosure, one of several located at the back of the line of sheds. We prevailed upon Mutsu-san to sacrifice his virtually unused golf practice net to string up over the enclosure, and then sowed grass seed and transplanted a couple of small trees into it. By July the grass covered the ground in a lush carpet, and we were ready to proclaim the "cat garden" open. On the morning of July 4, Chiyoko pulled back the small sliding hatch leading to the garden while I stood outside and called.

A few of the cats had had previous experience of the great outdoors, but for the majority this moment was their first encounter, albeit a limited one, with Mother Nature. All of the first group of cats to fill the entrance to the garden belonged to the uninitiated, and they were predictably stupefied by the sight that greeted them. Fearful of stepping out into the waist-high grass (their waists, not ours) and of the open sky overhead, but too curious to turn back, they just sat there paralyzed. I don't think that I shall ever forget the expressions on their faces. However, with others pushing them from behind, they were not allowed to remain there for long. Within five minutes almost every one of the cats (there were over sixty by that time) was out and stepping tentatively through the grass. Only the initiated, such as Pe and Ganko, started chewing blades of grass appreciatively as soon as they entered the garden, but within a short space of time others were joining them. Once over their initial astonishment, the toms regained their composure by industriously spraying on every upright surface, while kittens were soon darting up and down the trees and ambushing each other in the grass. Some adults, too, began to join in the fun, and a party atmosphere reigned.

The garden was in every way a great success. Only a hundred square yards, it was

I amassed an overabundance of snaps of luxuriously sunbathing, sleeping cats.

by no means large, but it was much better than nothing. Because the net would break under the weight of heavy snow, we had to remove it each winter, before the first blizzards arrived, and close the garden until spring, but for eight months of each year the cats were able to enjoy the feel of grass beneath their feet, the sky directly above their heads, and even rain on their backs if they so desired. In addition to sampling the joys of mating, fighting, playing and catnapping in the open air, they could work off their somewhat frustrated hunting appetites on an assortment of flies and beetles, even the occasional frog or shrew unfortunate enough to venture into the garden unawares. In the summer we would bring back from the forests bundles of vines of the species that the Japanese call *matatabi*, which has the same inebriative effect on cats as catnip; we would hang them from the branches of the trees and get all the cats good and sozzled. They could also join noses and exchange greetings through the netting with the horses, pigs and sheep that passed by outside each day. In brief, despite its shortcomings, the garden provided the cats with a whole host of new sensations and stimuli, and a chance to appreciate more directly the passage of the seasons. They showed their gratitude for it by gathering at the hatch, begging me to open it, as soon as I entered the building early each morning. One after another they would file out to sniff the air and chew the grass. I never tired of this somewhat incongruous sight, reminiscent more than anything else of a herd of cows set out to pasture.

Stupefaction on opening day of the garden.

January 10, 1980. We had the first heavy snow of the winter last night, and it took me twenty minutes to dig my way to the cattery entrance this morning. As I approached the door, I could already tell that something was different inside: the chorus of almost continuous monotone calls echoing through the building, unmistakably the cries of the toms searching for a female in heat. Sure enough, when I finally got inside, I found a number of toms who were distinctly agitated. When I scattered *niboshi** over the floor, they showed only passing interest in eating, paying far more attention to the females who had gathered in the area. It seemed that they still had not found their quarry. About a quarter of an hour later, I saw seven or eight toms gathered at one of the sawdust tubs, all sniffing intently at a certain spot and "flehming." Whoever the female was, she had clearly urinated there recently. It looks like the season of love has arrived, along with the snows.

Sha "flehming" in his special way.

It was the same every year. With the coming of really cold weather, the cattery would be quieter every day, with more and more cats spending more and more time snuggling up to each other in the winter beds I had prepared for them. Particularly on cloudy days, the building would look deserted between the daytime hours of ten and three, and for most of the night. Then, one day, invariably around the middle of January, the atmosphere would suddenly change. The sheds would reverberate with the insistent calls of the tomcats, usually all moving in different directions but all with the same purpose in mind—to seek out and mate with the owner of the particularly enticing smell that increasing numbers of them had found in one, or even two or three, of the sawdust tubs since the previous evening.

To identify and analyze this smell, the toms would have utilized not only their noses but also, through a behavior pattern called flehming (from the German *flehmen*, a word which has no literal translation), a

**Niboshi*, which literally means "boiled and dried," are sprats no longer than a couple of inches, which have been boiled to remove their oil and then dried. The cats loved them, and got a helping every morning—although only a small one because *niboshi* are not cheap.

Kerompa gives her suitors the cold shoulder between matings.

tiny pouch of specialized olfactory cells situated in the roofs of their mouths and known as the Jacobson's organ, after its discoverer. Apart from the Felidae, both deer and horses possess functioning Jacobson's organs and accordingly perform characteristic flehming. Anyone who has lived with horses is probably familiar with the grimace that a stallion produces by rolling back its upper lip and wrinkling its nose in response to the smell of an estrous mare. Although a cat also pulls its upper lip up slightly, its expression during flehming is not nearly as extreme as that of a horse. It would appear that the cat traps airborne odor particles on the taste buds of its tongue and presses these to the particular region of the roof of the mouth where a tiny duct leads to the Jacobson's organ, which could thus be viewed almost as a specialized "nose" within the mouth. The toms do not have a monopoly on this behavior, nor is the urine of an estrous female the only stimulus to elicit it. Females also flehm occasionally, and both sexes show the response to a wide variety of smells which, at least to us humans, seem to bear little relation to each other. There is no question, however, that the smell of the hindquarters of an estrous female, or of her urine, elicits more flehming on the part of the toms than any other smell.

It was no easy task for the toms to identify a female in heat. For one thing, their powers of olfactory discrimination were no doubt severely tested by the number of possible candidates—over sixty in the 1982 season, when the population of the cattery was at its peak. Moreover, the female concerned would most likely be buried among a whole host of others in one of the winter beds. Even if she were aware of the physiological changes going on inside her, she would still not be feeling even the slightest inclination to mate and would not know that she was the cause of all the consternation among the toms. Consequently, it might take the toms a day, sometimes more, to zero in on her. At first only one or two would discover her, but their pestering would be noticed by other toms, and very quickly she would gather a host of suitors.

Far from being flattered, a female in the first stages of estrus would find all this attention most tiresome, and would show her irritation in no uncertain fashion, by turning to hiss and flash her claws at any tom presumptuous enough to approach too close. The ease and authority with which a female could fend off even the most ungentlemanly suitors, and the tactics she used, never failed to impress me. When resistance down on the ground or within the winter beds became impossible, she would head up to a narrow perch near the ceiling, a maneuver which severely hindered the toms' advances. Despite the cold, she would stay there for hours at a time, surrounded on all sides by the toms gathered at sites just below her. Every so often the toms would utter characteristic abbreviated "chirrups," as much as to say, "How about it?"; but try as they might, the female would remain impervious to their demands until, eventually, she felt a little more in the mood for mating. Even then, I got the impression in many cases that, far from being eager to mate, she was at first merely condescending to let herself be mounted. It would still take some hours before she really began to get in the mood and actually solicit attention.

The sexual act in cats begins with the tom gripping the nape of the female's neck from behind, and straddling her. This nape grip has the effect of partially immobilizing a receptive female; she becomes almost as passive as a kitten being carried in a similar nape grip by its mother. Both the copulatory bite of a tomcat and the carrying bite of a mother cat are orientated in the same way as the bite with which a cat kills its prey, but differ from the killing bite in that they are inhibited, the cat concerned finding it literally impossible to close its jaws completely. Even so, the copulatory bite is by no means merely a loose grip. Indeed, if the female struggles, the tom is likely to grip all the harder, and occasionally causes her pain and light injury. The receptive female, however, rarely struggles, and invariably assumes the copulatory pose immediately upon being gripped—flattening her back, raising her rump and turning her tail to one side. If she does not do this quickly enough, the tom will stimulate her to adopt the pose by treading on her hindquarters alternately with his hind feet.

Having settled strategically behind Pe, Mo, who is not really asleep, waits for her to wake up . . . When she does, Mo jumps onto the back of the sofa for safety's sake, and inquires as to her mood . . . Pe doesn't exactly say yes, but neither does she say no . . . and so Mo takes his chance . . . Both cats have a wealth of experience, and the act proceeds quickly, since they are undisturbed by others . . . Having reached her climax, Pe goes for Mo . . . Fifteen seconds later, Mo, although eager to groom himself, can't tear his eyes from Pe, who is still in the throes of ecstasy.

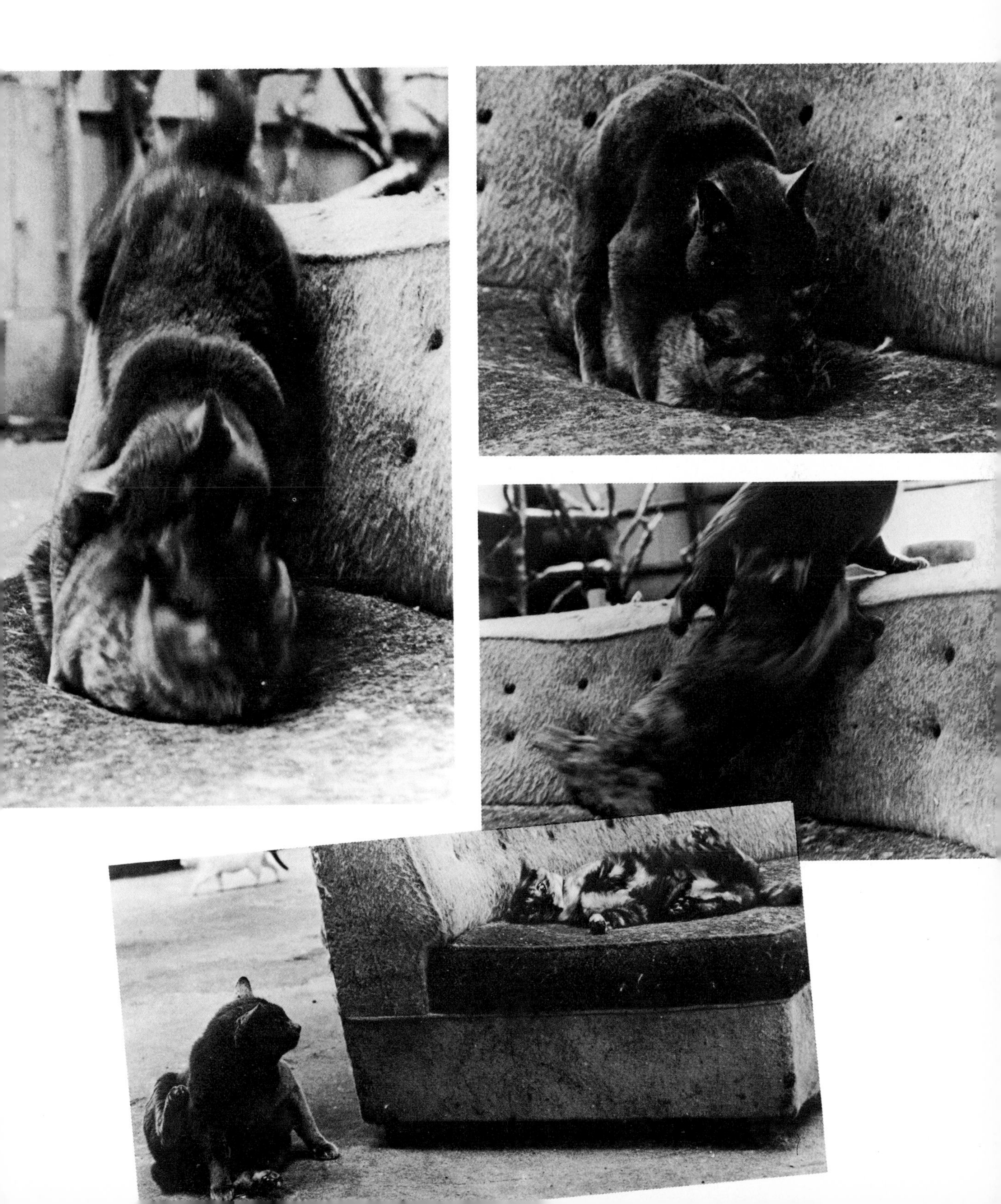

When she is ready, he stretches his torso backward to bring his penis within reach of the female's vulva. The rubbing of his genitals on the female's hindquarters often causes her to raise her rump further, making things easier for her partner. An experienced pair can accomplish intromission within about ten seconds of the male mounting the female, but, especially where an inexperienced tom or a not entirely cooperative female is involved, the process may take much longer.

Once intromission is effected, however, the act is over in no time at all, and is most dramatic. As soon as the two cats are joined, the female starts to growl and kick backward alternately with her hind legs, and although this is the only movement she is really capable of in her semiparalyzed state, it can occasionally result in dislodging the penis, especially of an inexperienced tom. A tomcat's penis is quite short, even when erect, and so its position within the female is not very secure. Perhaps for this reason, very few thrusting movements are required of the tom, who seems to be designed to begin ejaculating almost immediately upon entering the female. His penis is covered in short, posteriorly protruding barbs, the friction of which on the vaginal walls of the female is assumed to play an important part in achieving orgasm in both cats. I wouldn't mind betting, however, that an equally vital role of the barbs is to keep the penis in position when the female starts to kick. There is no rule in biology which states that an organ, appendage or receptor must have only one function; indeed nature is very economically minded, and multiple functions are the norm.

Within a matter of seconds of intromission, the female's growls escalate into ear-piercing howls; when it seems that they cannot get any louder, she twists her body, and any wise tom immediately leaps off her and adopts a somewhat defensive posture, ready to retreat further should the female strike out at him with her claws, as she is prone to do. At the same time, however, he seems almost totally transfixed by the sight of the female, who, if she hasn't turned on him, will now be alternately and furiously rolling and rubbing herself on the ground and licking her hindquarters. It is impossible to say exactly what thoughts are passing through the head of either cat at this time, but the female appears, to my eyes at least, to be in the throes of ecstasy, and it is conceivable that the tom derives pleasure from watching that ecstasy. Anyway, it is a sight that he won't miss at any price, and it is only after the female has quieted down a little that he attends to licking himself. Even then, he does not let the female out of his sight, and drops everything to race after her if she decides to change her location.

I have described how the act of copulation should proceed under ideal conditions, with only one tom courting the female. In the cattery, however, a popular female might at times be surrounded by more than fifteen toms, and in such situations the act rarely went so smoothly. To be sure, not all of the toms escorting a female were equally eager to mate. For example, out of ten suitors, only four might at any one time attempt to mount her, the remainder always amazing us by their apparent willingness to sit on the sidelines and merely spectate. Even with only four really earnest suitors, however, the female would often suffer considerable frustration and also, no doubt, some physical pain, as all four of the toms pulled her nape this way and that while jockeying for position. The female might have to contend with their jostling for more than fifteen minutes until one tom finally effected intromission. The fact that many a female lost most of the fur on the nape of her neck, which in addition frequently became swollen and sore, is testament to the rough treatment that they underwent during estrus.

Even if the toms did not suffer physically, as a result of their numbers they were subject to as much mental frustration as the females, if not more. A tom would mount a female only to relinquish his position a few seconds later when another tom also latched onto the female's nape. Sometimes a tom would dismount even without suffering interference, the mere presence of other toms around him appearing to put him off. As a result, a tom who might have mated up to ten times in a one-to-one situation would be lucky to succeed only once or twice.

Needless to say, some of the toms adapted better to these adverse conditions than others, and none better than Nibu. This cat seemed to be blessed with an excellent nose, for he was invariably the first tom to sniff out a female approaching estrus. He also invariably led the vanguard of toms in whittling down the resistance of such females, and as a result was frequently on the receiving end of painful paw blows. This never worried him overmuch, however; "pushy" is too polite a word to describe him. He was, and from all I hear, still is, the most ungentlemanly tom in the cattery, even going so far at times as to bully a female who resisted his advances.

His determination produced results, for he was more often than not the first tom to mount and mate with a female to whom he had taken a fancy. Once on top, unlike most of the other toms, he was almost impossible to budge. Moreover, during the 1979 season he developed a very effective strategy, which has served him well to this day, for dislodging other toms who occasionally beat him to a female. Approaching from behind, he would push his head between the mating couple and, when far enough forward, he would hump his back and cause his unfortunate rival to topple to the side. Though a simple enough technique, very few of the other toms ever mastered it; cats, although tireless observers of each other's behavior, are apparently not great copycats.

As a consequence of his determination and skill, Nibu would often monopolize a female while he remained interested in her. Wanchan, however, represented the other extreme. His mating performance was quite good during the first season in the cattery; but he also developed a habit which was

Two toms latched onto the nape of a female's neck, while the third tom nosing in from behind is Nibu, demonstrating his special technique for dislodging other toms.

as counterproductive as Nibu's was productive, and quite comical as a result. Eager though he always seemed to be to mate, he became increasingly hesitant about mounting females, instead standing on his hind legs with his forepaws pressing into the back of any tom straddling the female. He looked for all the world like a teacher instructing the tom below on the finer points of sexual technique. He would remain in this position even after the tom below had accomplished intromission, staring down intently at the female howling below the two of them. This somewhat aberrant behavior took increasing hold of his psyche, and by 1982 clearly dominated his sexual repertoire. It was almost certainly the vast excess of willing males in relation to the number of receptive females at any one time, a factor beyond Wanchan's control, which was at the root of his peculiar deviation, and as such, it was unfair of us to

make fun of him; but he inevitably became a major attraction of the mating season in the cattery.

While on the subject of peculiar behavior, I should also mention the occurrence of homosexuality in the cattery. I use the term purely descriptively, with reference to the mounting of one tom by another. It is extremely doubtful in most cases that the tom doing the mounting was actually aware that his partner was a tom like himself. If such superficially gay toms had been able to speak, they would no doubt have sworn that they were straddling a female. Moreover, the tom being mounted was in most cases a rather unwilling partner who suffered the insult only because he was as prone as any cat to a degree of paralysis upon being gripped by the nape of his neck. After a short while a mounted tom would begin to protest and roll onto his side to try to push off his assailant with his hind feet, but this rarely put off a determined tom who would invariably try to remount. I never once witnessed a mounted tom get angry with his assailant; mounting is clearly not aggression-eliciting behavior.

Most such behavior occurred among the toms surrounding a receptive female, usually while they were waiting for her to consent to mate, and most of the perpetrators were young toms with limited sexual experience, although older toms were occasionally involved. In a sexual tizzy, so to

Nibu gets punished for his pushiness by Mike-modoki.

speak, such toms appeared to have lost their powers of discrimination, and so, when their path to satisfaction was blocked by the uncooperative attitude of a female or by the monopolization of the female by other toms, they tended to redirect their attention to the nearest reasonable substitute.

The same goes for those toms who on occasions actually seemed to enjoy being mounted and actively solicited such behavior from other toms. Most such soliciting was shown by toms in a state of sexual excitement: by adopting the female role, they were merely giving unconscious expression to the bisexual disposition which exists in many mammals, including man. Moreover, it is not only male animals that show such behavior. I had been used to seeing females among our dogs attempting to mount each other when in heat. I knew also that cows were notorious for such antics. And so when, one day in the spring of 1981, I saw a young estrous female in the cattery turn on Mo, one of the crowd trailing her, and mount him, gripping his nape between her teeth for about five seconds, I was not very surprised by her behavior as such. What did surprise me was its appearance in the cattery where females were never lacking in suitors. The female concerned certainly wasn't, since there were about eight other toms chasing her, apart from Mo.

Perhaps one of the strangest incidents occurred during the 1981 season. Pe, who was then at the height of her first estrus of the year, had mated about five minutes before and was walking along one of the aerial walkways in the far shed toward the ladder which slanted gently down to the passage joining the two sheds. She was not yet in the mood to mate again and was swearing

One tom astride another, who doesn't seem too happy about it.

Howling abuse at each other at point-blank range.

A weak opponent expresses at one and the same time both his inferiority and his resolve to defend himself.

Sha shows his impressive fangs in an altercation with Wanchan.

Claws flash and fur flies as Panda pushes Hi to the edge of the hammock.

Ganko under pressure from Peyo. Only the weakest toms turn and flee.

Peyo now threatens Ganko broadside.

Two females engaged in a duel. All female duels were very one-sided.

A very confident Kurokasan glares at a very scared Uko. Note the difference of ear angle. Kurokasan is enjoying herself.

The garden entrance, a notorious blind spot, was the scene of many confrontations. Here brothers Jerry and Hi face up to each other.

warningly at any tom who came too close to her.

Just at that moment, Cherry, a young immature tom of about seven months, meowed somewhat anxiously from the top of the far shed shutter. Cats frequently climbed up there only to find, once they had finished their investigations, that they had trouble getting back down. They had either to leap down about four feet to the ladder, or horizontally about the same distance to the aerial walkway along which Pe had been proceeding. Although this was not a difficult proposition, young cats unfamiliar with the location often displayed a lot of hesitation before finally making the leap. Cherry was one of Pe's kittens from the previous summer, and as soon as she took in his predicament, she seemed to forget all about the toms around her. She stopped halfway down the ladder, looked up at Cherry and answered his anxious cries with the unmistakable calls of a worried mother. This astounded me. She had not shown the slightest signs of maternal behavior toward her kittens for over three months, and then suddenly, slap-bang in the middle of her heat, she produced this performance, showing her concern for the full half minute it took Cherry to pluck up enough courage to leap over to the walkway. During this time the toms, almost as if they understood and respected Pe's feelings, desisted from making advances.

Pe clearly still recognized Cherry as her child, although usually she no longer paid him and his siblings more attention than she did any other cat. I have no doubt that Cherry also was aware that it was his mother answering his calls so earnestly, but as he had long since become independent of her, her concern did not appear to mean much to him. Once he had reached the walkway, he wandered off in the opposite direction. Far more surprising, however, than Pe's recognition of Cherry, despite the number of cats in the building, was that she suddenly chose this particular time to show maternal concern. In the cattery, the opposite was usually the rule. Females would invariably begin to reject the approaches of their kittens just before the onset of another estrus, about two to three months after giving birth. Pe's behavior was exceptional, and I still have no explanation for it. The timing of the encounter was, however, significant, falling as it did just at that point, shortly after intercourse, when Pe was still in no mood for the next mating. With her sexual appetite temporarily sated, the high level of estrogen circulating in her system might have rendered her particularly sensitive to stimuli eliciting maternal behavior.

Toms are usually fickle lovers, but then it would not pay them to spend too long with one female.

6

FICKLE LOVERS

With all this frantic activity, the cattery during the mating season was lively, to say the least. Anarchy and "free love" appeared to be the order of the day. But, looking below the surface, there were in fact definite patterns to be detected in the sexual activities of the inhabitants. The chart on page 63 is a record of Pe's estrus in early 1980, and shows which males displayed interest in her, the approximate period for which that interest lasted, and the number of successful matings witnessed. It is by no means a complete record, as I was not watching her for twenty-four hours a day, but from my observations of the course of countless estrous periods, I can vouch that it is fairly typical for a popular female, and reflects the general features reasonably accurately.

Not every tom mated, or even courted, every estrous female. Pe was quite a favorite with the toms, but even she could attract only nineteen of the twenty-seven mature toms present in the cattery at that time. Her brother Sha ignored her for some reason on this occasion, although he had mated with her on others. Another notable absentee was Ganko, who spent most of this period pursuing a young female called Choco. Mo almost missed the boat, eventually "discovering" Pe after spending two fruitless days pestering a very unresponsive Uko, his favorite from former days; it often happened that toms would waste time (from the point of view of realizing their reproductive potential) with favorite females of theirs, whether the latter were in heat or not. Other absentees during Pe's estrous were simply more interested in other females in estrous at the same time, as were many of Pe's suitors outside the time that they spent with her.

However, even when only a single female was in heat, she would be completely ignored by quite a number of the toms, demonstrating that tomcats, although their reputation as philanderers is well deserved, are not totally lacking in partiality when it comes to choosing mates. In more natural environments it is likely that other factors, such as the distribution of territories among rival toms, affect mate selection more than the matter of personal likes and dislikes; even in Pe's case, this was true to a certain extent. Waru wanted to mate with Pe, an old favorite of his, from the second day but, well aware of his unpopularity among the toms, he hesitated to come too close to her while she was surrounded by so many of them. He was able to mate with her in the latter part of the third day, but when Mo discovered her on the fourth day, Waru had no choice but to keep his distance.

Such rivalries in the cattery affected the mating behavior of only a small portion of the toms. Most of them were on good enough terms with, or indifferent enough toward, each other to be able to concentrate on the female they were after without taking much notice of each other's presence. Many also showed a partiality for certain periods in any receptive female's estrous: some preferred to mate with her in the first two days of her estrous, while others found her more attractive a little later on. As I have already mentioned, Nibu was an extreme example of the former type, invariably going only for females in the first stages of estrous, and the chart reveals that the majority of toms followed him in his preference.

Females were not always the most willing partners in the sexual act during the first couple of days' mating. Around the third day, they would finally begin to show a growing enthusiasm for mating, which was paradoxically accompanied by an increasing lack of interest on the part of the toms. A female in the last stages of estrous was often a pathetic sight, seemingly having to pull out all the stops to persuade a last remaining, somewhat halfhearted suitor of her charms. She would often literally tremble with desire as she went through her repertoire of enticement, rubbing herself

CHART SHOWING PE'S SUITORS, THE PERIOD THAT EACH TOM SHOWED INTEREST IN HER AND THE NUMBER OF SUCCESSFUL COPULATIONS OBSERVED

SUITOR / DAY OF MATING	DAY 1	DAY 2	DAY 3	DAY 4	DAY 5
Nibu	- - - - • • • • -				
Panda	- - - - - • - - -	• • - • - -			
Sabu	- - - - • - • - -				
Wanchan	- - - - - - - - -	- - - • - - - - -			
Kenjiro	- - - - - - - - -	- - - - • • - - -			
Hige	- - - - - - - - -	- • - - - - • - -			
Chiro	- - - - - -	- - • - • - - - -			
Mambo	- - - - - - -	- - - - - - - - • -			
Chobikuro		- • - • - • • - -	• - - - -		
Kenbo		- - - - - - - • -	- • • - - - - • - -		
Cocoa		- - - - - - - - -			
Marshmallow		- - - - - - - - -	- • - • - -		
Osato		- - - - •	- - - -		
Ee		- - • • -	- • - - - - - - -		
Waru		- - - - -	- - - - - • • - •	- - - - - -	
Hi		- - - • -	- • • - - -		
So			- - - - - • - • -	- - - - - -	
Mo				- • - • - • • - -	
Otochan				- - - - - - - - • •	- • • - -
TOTAL SUITORS PER DAY	8	14	8	4	1
TOTAL WITNESSED MATINGS PER DAY	7	20	14	6	2

- - - - - - denotes period of interest in Pe shown by each tom.
• denotes observed successful copulation.

against the tom, pushing her hindquarters into his face, adopting without the slightest encouragement the copulatory pose to the full extent of raising her rump, holding her tail to one side and treading with her hind feet, and then looking back at the tom and chirping beseechingly in his direction. Such a performance would end in failure as often as in success. It always struck me as ironic that, although there were never enough estrous females at any one time to satisfy the sexual appetites of all the toms, such blatant soliciting by a female often went unanswered. To the human observer, the toms appeared distinctly malicious in their preference for pestering unwilling females while turning a blind eye to those offering themselves, so to speak, on a plate.

What could be the reason for this lack of synchronism in the sexual moods of the toms and the females? Professor Paul Leyhausen, in his authoritative book, *Cat Behaviour*, notes that this phenomenon occurs not only in all members of the cat family that he has observed but also in many other mammals, and so an explanation based on the behavior of domestic cats alone would be inappropriate. It seems possible to me that the answer lies in the amount of time inherently required by the hormonal processes governing the onset of estrus to take effect. While toms would be alerted through the smell of the female's urine to the changes starting to take place in her physiology, it would not pay the female to get in the mood for mating until these changes were nearing completion. Just as the effects of hormonal processes take time to build up, they also take time to wear off, and the female's continued desire to mate long after most males have lost interest in her might be the result of such momentum. It is possible, however, that this delay between the time that a female begins to leave signs of her approaching estrus and the time that she is willing to mate may be distinctly advantageous to females of solitary animal species. By unwittingly advertising her condition while still not in the mating mood, a female cat no doubt often saves herself the bother of having to search actively for a mate.

The record of Pe's estrus also reveals that no single tom spent longer than two days courting her. In my opinion the brevity of a tom's interest in a female is intimately connected with the method of ovulation found in cats. Cats are "induced," as opposed to "spontaneous," ovulators. This means that female cats require the physical stimulus of copulation or its equivalent with a substitute object in order for ova to be released. The domestic cat shares this requirement not only with its wild relatives but also with the rabbit, shrew, ferret and some other mustelids, while a host of other mammals, including the horse, cow, sheep, dog and ourselves, ovulate spontaneously at the peak of estrus. Induced ovulation could be of clear advantage to animals leading solitary life-styles, and I think it likely that it is the norm with many more mammals, particularly those of solitary habit. In fact, it is possible that induced ovulation is the more primitive process.

An estrous cat who is normally given access to males and mates is over her receptive period within about five days, if not sooner. If she is isolated, however, her estrus may last up to two weeks. It is in this prolongation of estrus in the absence of copulation that the advantage for solitary animals lies. Whereas the finding of a mate would not be problematic for animals liv-

Mike-modoki beseeches Ee to mount her by "chirping" and parading in front of him.

ing in groups, the same cannot be said of such solitary and sparsely distributed animals as the domestic cat's wild ancestors and relatives. Prolongation of estrus in the female would provide an equivalent increase in the chances of her being discovered by a male, and thus of conception.

I am inclined to think that the explosive reaction of the female during mating is fundamentally linked to this mechanism of induced ovulation. Her response of fierce howling is as consistent as it is dramatic, and so it should be in view of the fact that it is the prerequisite of ovulation. In other words, in induced ovulators, female orgasm is probably essential for ovulation, and thus for success in reproduction, and females would accordingly be designed, as the cat appears to be, to achieve orgasm with comparative ease. If, indeed, induced ovulation is the more primitive mechanism, and if the prime purpose of female orgasm is to induce ovulation, we could expect the importance of female orgasm to be diminished in those species which ovulate spontaneously. There would still have to be something pleasurable in the sex act for females of such species, in order to ensure that they mated willingly, but a high-intensity response, like the howling orgasm of the female cat, might not be essential. This line of thought would seem to be substantiated by the remarkably staid response to the act of copulation displayed by estrous mares, cows, sows, sheep and does, all of which are spontaneous ovulators. Judging from outward appearances, the female cat seems to derive far more intense pleasure from copulation than any spontaneous ovulators with which I am familiar.

Induced ovulation in the cat may also have far-reaching implications for the mating behavior of toms. Because copulation stimulates ovulation, conception is virtually guaranteed even after a very limited number of matings. The male would therefore gain no particular advantage by remaining with the female throughout her estrous period. Once he had spent a day or two with her, and had mated several times, he might be well advised to move on, to turn his attention to other females in the area. The behavior of the cattery toms seemed to provide support for this supposition. As females were never short of mates, and copulated as soon as they came into estrous condition, their estrous periods lasted on average about four or five days; but, as mentioned earlier, no single tom maintained an interest in a receptive female for longer than two days, no matter at what stage in her estrus he first began to court her. No doubt in some cases, such waning of interest was hastened by the appearance of estrus in other females, but even in the absence of other estrous females, toms displayed the same loss of interest, particularly after mating a few times. In Pe's case, for example, both Nibu and Sabu, who had virtually monopolized her on the first day of mating, seemed no longer to give a damn about her when I watched them early on the next morning. Although there were no other females in estrus, both these toms were clearly on the lookout for likely females, and were calling incessantly. Yet when I brought Nibu to Pe and placed him right behind her hindquarters, he did nothing more than cursorily sniff her vulva, and then went on his way. I tried the same thing on other occasions with other toms and the result was always the same. Once a tom had tired of a particular female, no amount of persuasion could get him to mate with her again.

Of course, induced ovulation may not be the only factor determining such an attitude. For example, high kitten mortality would favor a strategy on the part of the toms of roaming broadly, meeting and courting as many females as possible. The fact that most toms, both in the cattery and in free-range situations, are irrepressible philanderers suggests that the rewards of such a strategy exceed those of remaining with a certain female and assisting in the rearing, protection of and provision for her kittens, even if more kittens would survive as a result of such paternal behavior.

In any case, it would be difficult for a tom to stay with a female twenty-four hours a day for the duration of her estrus. While he could follow her when she wandered off to hunt, he would probably have trouble persuading her to come with him when he began to feel peckish. In his absence the female would be free, if she felt so inclined, to mate with other boyfriends. And even if the tom were able to remain with her, it is doubtful whether he would be able to keep rivals from getting at her, particularly if

Six toms wait peacefully for Nougat, the tortoiseshell and white in the center, to signal her readiness to mate again. One, to the right, is a little impatient and tries mounting another. The posture of Panda, waiting above, is eloquent.

they were numerous. In the cattery, fighting in the immediate vicinity of an estrous female was astoundingly rare, and even when it did occur, it tended to remain localized and almost halfhearted. Even as two toms stood glaring and growling at one another, one could almost sense their unwillingness to prolong the duel and their eagerness to get back to the matter foremost in their thoughts, the estrous female. And well they might, for while they faced up to each other, the other toms would be virtually ignoring them, so intense was their concentration on the female. The more intolerant a tom was, the more likely he was to damage his own chances of mating.

Almost as if the toms realized this, they displayed in general a remarkable tolerance toward each other while they were near an estrous female. Their familiarity with each other, together with the tendency of particularly bitter rivals to avoid pursuing the same females at the same time, was no doubt another factor adding to the relative peacefulness of the situation. However, something in the expressions of the courting toms made it difficult for me to accept this explanation *in toto*—particularly since, outside the immediate vicinity of receptive females, they tended to fight more rather than less frequently during the mating season. It seemed to me that the total absorption of most of the toms in the female had the effect of inhibiting their aggressive impulses. They seemed almost totally oblivious of each other, treating each other not as rivals, but merely obstacles to be overtaken, dodged, jostled, dislodged or, in the case of homosexual mountings, as sexual substitutes. The estrous female seemed to blind the toms to each other as rivals, and it appeared that they had to escape the spell that she cast in order to have a good brawl. When considered in conjunction with their tendency to lose interest in a female with whom they had mated a few times, this lack of aggression would appear to make even more sense. After all, there is little logic in fighting fiercely for a sweetheart one day, if one is going to leave her to some other tom or toms the next.

These arguments could apply equally well to the whole spectrum of free-range situations, but I would not expect free-ranging toms to show the same tendencies to the same degree. Particularly in rural environments, with relatively sparse cat populations, territorial considerations would probably play a major part in deciding who mated with whom. There are limits to the distance which a tom can wander in search of females, and thus to the number of females he can court. If he has nothing better to do, he may well remain with a female even after he has lost interest in her sexually (especially if, as often happens, he shares part of his range with her), and by his mere presence prevent the approach of other toms. In large towns with dense cat populations, however, the territories of toms show a great deal of overlap, and many of the toms in a particular neighborhood might be on reasonably familiar, if not licking, terms with each other. In such a situation we might be able to observe scenes among courting toms reminiscent of those I have described for the cattery. Moreover, as the number of females in a given area would be much greater than in a rural environment, town toms could be expected to be busier than their rural counterparts in their courting, and show the same rapid turnover of females witnessed in the cattery.

One area in which my observations differ from those made by others is that of the female cat's ability to choose her mates. Leyhausen states that in both caged and free-range situations, "choice of partner is something which is almost always decided by the female," offering this as a reason why the dominant tom cannot monopolize all the females in the area and exclude other toms from mating.* I can easily imagine that a female could exercise choice if the number of her suitors was limited to two or three, and she would no doubt be far more hesitant about inviting the advances of a tom whom she did not know than of a tom she knew and had mated with before. In the cattery, too, certain females had their favorites among the toms, often associating with them more than with other toms outside their estrous periods. Length of acquaintance rather than rank appeared to be the deciding factor in the friendships; for example, the dominant cattery tom, Mo, enjoyed warm relationships with Uko, Aya and his sister May, all females with whom he had passed years in the main house, but none of the other females paid him any special attention, or he them, outside their estrous periods. When it came to mating, however, such friendships appeared to be as irrelevant as rank in determining who copulated with whom. Once she had con-

*_Cat Behaviour_ (New York: Garland Publishing, Inc.).

sented to the act, a female had little choice but to accept any tom who mounted her, such were their numbers. And after the usual prolonged jostling and jockeying by the toms, the female was no doubt heartily grateful for any tom who managed to give her satisfaction. At least where the cattery toms were concerned, mating success depended far more upon individual determination and sexual expertise (with Nibu as the prime example) than on either rank of the toms or preferences of the females.

The cattery can also provide answers to certain questions which have up to now remained open. According to Leyhausen, one such question is "whether a male courts several females at once or only one at a time." Since, during the mating season in the cattery, there were frequently more than two females—occasionally as many as seven—in estrus at any one time, the toms had ample chance to court several females simultaneously, but I only once saw a tom avail himself of such opportunities. Panda, the tom concerned, was, with five or six other toms, attending a female called Nougat who was in the second day of her estrus (i.e., the day after she had first mated). Panda was a very impatient tom, and when the female of his fancy was snoozing or otherwise unamenable to advances, it was his habit to go wandering around the building, spraying here and there and calling for females.

On that day, Panda went off for a wander while Nougat was taking a nap between matings. On his way, he came across Mike-modoki, a female with whom he had mated a few times on the previous day. She was well into her estrus, and earnestly soliciting the attentions of any tom she spotted, but with little success. Panda let himself be tempted, although one could almost tell from his attitude that his heart was elsewhere. He completed the act perfunctorily, with a minimum of bother, and when it was over, instead of remaining with Mike-modoki to watch her postcopulatory rolling and licking, he almost immediately took himself off to check on Nougat again. While it is more than likely that other toms also occasionally showed such opportunism, as a rule the cattery toms only showed a sexual interest in one female at any one time.

The feature that impressed me most about mating in the cattery was its egalitarianism. Even if rivalries did play some small part in determining the composition of the group of toms around an estrous female at any given time, in general their effect was insignificant. Even the strongest toms showed no inclination to monopolize females by force, and almost all toms were free to mate with whom they wanted, the only important conditions for success being determination and sexual expertise. This is in sharp contrast to many animals living naturally in groups, when rival males are often banished from the group, or take no part in mating activity which is monopolized by the strongest of them. While strong toms in a free-range situation might be able to secure larger territories and accordingly court more females, they seem unable to use their rank to much effect in a group situation.

7

MORE BLUSTER THAN BLOODSHED

"Why do your cats fight so much? Even if they are confined to those two sheds, they still have plenty of space to exercise and play, plenty of food, warmth and company. What do they have to fight over so much?"

This was the question posed to me one day by a visitor to the Kingdom who had witnessed three tomcat duels during the half hour that he had spent in the cattery. As this particular visitor was a lecturer at one of Japan's most respected veterinary colleges, I was somewhat surprised that he seemed genuinely puzzled by our toms' aggressive behavior, but then persons far more qualified, such as psychologists and sociologists, have repeatedly fallen into the trap of attempting to explain all aggression as merely a reaction to certain external circumstances. Implicit in the lecturer's query was the suggestion that there was something, he knew not what, in the living conditions of the cats which was causing them to fight. If so, by making certain corrections to the environment, I would theoretically be able to put an end to the fighting. I explained to my guest that the toms would continue to fight each other even in Paradise, and that if I were able to release all the cats into a far larger area, I might significantly reduce the number of fights, but those that did occur would almost certainly be much more prolonged and violent.

Fights in the cattery rarely lasted longer than five minutes and were almost always concluded without any physical violence, through a tense but harmless exchange of threatening postures and vocalizations. In my present location in a residential area of a Tokyo suburb, however, I have been an eyewitness to many confrontations (and have been kept awake by many more!) each lasting more than half an hour, during which time the combatants may clash violently three or four times. I also have vivid recollections of the two occasions when, still a newcomer to the Kingdom, I forgot to close a door behind me and thus enabled two of the three mature toms present at that time to cross swords. One of the toms was Mo, the other a big, old, black tom who, by virtue of two white stripes, one across his chest and one across his abdomen, was called Brapants. With only a door separating them, these two toms were well aware of each other's existence and were always trying to get at each other. When my carelessness allowed them to fulfill this dream, they dispensed entirely with threat displays, going for each other tooth and claw from the start, with a fury which made the everyday duels of the cattery look like innocent party games by comparison.

If aggression is merely a reaction to certain external stimuli, how are we to account for the variation of aggressive response to the stimulus presented by another tom cat in the cases described above? Clearly something more than mere "reaction" is involved, something which suggests the existence of an appetite for aggression which builds up in the absence of an appropriate outlet. In other words, aggression would appear to be an internally generated drive which differs in no fundamental way in its spontaneity from the instinctive drives of living creatures to eat or make love. In the same way that the sight of a fully laden table will delight the eye and prod into action the salivary glands of a hungry person far more than it will those of someone who has just got through a three-course meal, so the sight of another tomcat will elicit a far more aggressive response from a tom who has been kept in solitary confinement than from one who daily meets his foes.

While it is not hard for the average human being to appreciate the species-preserving function of the drive to eat or make love, the war-torn history of his race and the dilemma it faces today make him justifiably doubtful of ascribing any such species-preserving function to the aggressive drive; but the fact is that, among a host of

Mo and Sha exchange insults.

other functions, aggression, or the threat of aggression, is the motive force behind the regulation of density and distribution of a population of animals in relation to available resources. As such it is as vital to the preservation of a species as the drive to eat or reproduce. It is through intraspecific aggression that territories are established and hierarchies created, and it is that threat of aggression that encourages most animals most of the time to respect boundaries and rules of social conduct.

Fighting, however, does not necessarily mean killing. Any species whose members consistently fought each other to the death would exterminate itself. Clearly, the purpose of intraspecific aggression is not to kill one's foe, but to impress upon him one's superiority, and in so doing either put him in his place or get him out of one's hair. Thus we find that animals have evolved various mechanisms for settling their disputes with the minimum of injury to either party. The better equipped they are for killing, the more necessary become behavioral mechanisms for minimizing the danger of inflicting injury upon those of their own kind and thus we find that ritualized threat displays, specialized fighting procedures and aggression-inhibiting appeasement gestures are most developed in predatory species.

Possessing four sets of needle-sharp claws in addition to a powerful jaw containing four equally sharp canines, the cat is a very well armed animal indeed, and it is hardly surprising that it has evolved an elaborate threat display consisting of very stylized postures, facial expressions and vocalizations, the performance of which enables combatants to measure each other's strength and resolve and, far more often than not, settle a dispute without any physical contact whatsoever. It is true that first-time adversaries will invariably exchange bites and blows in addition to curses, as will toms kept in isolation, due to an accumulation of their aggressive drives. However, most tomcat duels are characterized by extreme caution on the part of the combatants and strict adherence to an unwritten law which would appear to state that the participants must do all in their power to settle their differences verbally, with physical exchanges to be considered only as a last resort. While I dislike as much as any other cat lover the sight of two cats trying to tear each other to shreds, I must confess to being an enthusiastic fan of tomcat duels. There is, after all, a lot to admire in a mechanism which so effectively minimizes physical injury, but quite apart from this, there is, to my eyes, drama, tension and even beauty in tomcat duels to match any of nature's spectacles. I must have witnessed well over a thousand duels a year in the cattery and yet I never tired of watching one more.

A strong and confident tom bearing down on an opponent can present a magnificent sight. He straightens and stiffens his legs and approaches with slow and measured paces, not directly at his foe but slightly broadside, as if he is attempting to impress his foe with his physique. He also holds his tail stiffly in a kind of inverted capital L shape, the proximal quarter held horizontally in the line of the backbone, with the distal three-quarters pointing groundward (this, of course, applies to toms possessing a full-length tail, which is not very common among Japanese cats). The pupils of the eyes of a really confident combatant show little dilation, and the ears are pricked up and turned slightly backward, twisted outward at their base in such a way that the rear surfaces face forward.

Having approached to within about two yards of his opponent, the aggressor turns to approach directly. His pace may now drop even further. As he takes a step forward he is likely to swing his head to one side, holding it in that position until he takes another step forward, simultaneously swinging his head in the opposite direction. Despite this head swinging, he tries to keep his eyes fixed on his opponent. If he has not started to do so already, he is likely at this stage to start vocalizing in a kind of menacing hybrid of a growl and a howl, although the timing and pitch of such utterances are very much dependent on the degree of his confidence and on the response of his opponent. Nervous excitement may cause him to salivate profusely, with the result that his howls occasionally degenerate at their tail end into a kind of gurgle as he swallows to rid his mouth of saliva.

Having approached to within a yard of his opponent, any but the most confident tom transfers his center of gravity to his hindquarters, lowering them by bending his hind legs. This posture gives him the leverage to spring at his opponent, but it may equally be considered defensive, since he is now better able to raise a front paw in response to the threat of a counterattack. The closer he gets to his foe, the more such defensive elements appear in his threat display. For example, he tends to raise his head, up to now held low on an outstretched neck, and tuck his chin in as a further precaution against counterattack. Another action invariably seen in cats threatening at close quarters is the flicking of the tip of the tail to right and left; this may well be a feint movement designed to distract an opponent's attention. As this is just one more component of the instinctive actions comprising the threat display, even cats with only stubs of tails go through the motions of flicking them, although it serves them little practical purpose.

From this point a variety of actions may follow. The aggressor may suddenly leap for his opponent or he may remain in the same stance, continuing to threaten with howls and grimaces. If he has decided that he has made his point and chooses not to take the matter further, he will gradually cease threatening and walk away slowly at a tangent, keeping his eyes on his opponent, or he will sit on his haunches, look away or sniff the ground around him, allowing his opponent to retreat carefully. His behavior will depend as much on his resolve as on his opponent's response, since, in the event of a physical clash, even a weak opponent will defend furiously and can often deliver as good as he gets in the way of injury.

In a duel in which both participants are of roughly equal rank, the postures and vocalizations of one are likely to mirror those of the other in such a way that the two cats often approach each other until they are literally staring down each other's throat, or have their heads aligned, howling into each other's ears, both equally determined not to give way. I particularly remember Sha's duels with Panda as dramatic examples of such razor-edge confrontations. Sha possessed the most magnificent set of canines in the cattery, and on more than a few occasions I saw him bring those teeth very slowly within a millimeter of Panda's scalp as they both howled at each other at point-blank range. Even with whisker touching whisker, however, these two rivals rarely took matters further.

When there is a marked difference in rank between two dueling toms, the weaker party adopts an attitude which clearly indicates his inferiority but at the same time expresses his intention to defend himself with all his might should his aggressor decide to go through with an attack. He knows instinctively that to attempt escape would be to invite pursuit, that by staying put and defending himself he stands the best chance of weathering the threat and thwarting actual physical attack. Occasionally he does flee, heading for cover or any location from which he feels better able to defend himself. If he is lucky, his opponent will not pursue him, but in my experience most cats cannot resist the sight of a fleeing opponent and will give chase, eventually forcing the target to turn and defend himself. Habitual "fleers" often carry the marks of their cowardice in the form of scars on their hind legs, tails and rumps, proof that flight is not the best of strategies.

Assuming that a weak tom has not fled, on being approached by his foe in the slow, cautious manner described above, he crouches with his ears pointed backward and flattened to his head. Adrenaline discharge, triggered by his fear, causes the hair all over his body to stand on end and his pupils to dilate. With his opponent closing in, the hisses and growls with which he first responds to the threat escalate into piercing howls as his fear mounts.

Interestingly, the more his target screams, the less inclined the aggressor is to vocalize; he tends to approach silently, his silence only adding in the eyes of the human observer to his appearance of menace and malevolence. However, if anything, the aggressor grows more cautious the more extreme the defensive attitude of his target becomes. If his foe raises a paw in readiness to strike out defensively, he invariably halts his advance and raises one of his own forepaws in response. He no longer looks so sure of himself, especially when he sees his foe now rolling on his side, freeing first his forelegs and then his hind legs for use as a very effective shield against attack. A more confident foe, who does not adopt such an extreme defensive attitude, is in a way a much more approachable target. Faced with four sets of outstretched claws, almost any attacker is given pause for thought, and physical attack in this situation is as rare as, if not rarer than, it is in duels between more equal opponents.

Sha regularly got as close to Panda's scalp as this, but rarely went further.

Between this extreme and the opposite, when a cat answers an aggressive threat display with an equally aggressive one, lie all manner of variations: angle of ears, degree of crouch, vocalization, dilation of pupils, and so on, are all significant indicators of the rank of a tom in relation to his opponent.

Should the aggressor go through with his attack, the defender will at first attempt to fend him off with his shield of claws. Even a cat attacked in a standing position reacts with lightning speed to his opponent's leap and is usually able to roll in time to meet it in the same fashion as a cat defending from a lying position. If the attacker succeeds in penetrating the shield, the defender now switches tactics, pulling his attacker to himself with his fore claws while using his hind claws to do whatever damage he can to his foe's abdomen, and embedding his canine teeth in any region of enemy flesh available—usually elbows, shoulders, neck or forehead. His attacker is, of course, doing exactly the same. For a few seconds all hell breaks loose as the two combatants roll, buck, bite, kick, growl and send fur flying. Dogfights appear tame by comparison, and perhaps for this reason can go on much longer. Dogs, after all, have only their teeth as weapons. With claws as well to contend with, it is small wonder that cats do not make a habit of prolonged physical exchanges. When they separate they tend to spring apart as suddenly as they had fallen together, often taking up the same postures that they were in before the tussle. A weak tom will thus immediately roll over and once more stretch out his claws to fend off further attack. Unlike dogs, cats possess no specific appeasement gesture with which to inhibit attack, and a defeated tom, instead of offering himself to his opponent, has no choice but to continue defending himself. However, his determination to do just this is usually enough to avert a second attack in cases where it was not enough to avert the first.

It is invariably the higher-ranking combatant who leaves the area of battle first, since the inferior cat usually remains rigidly on the defense. In the cattery, other cats frequently stopped to sniff at a spot where a duel had occurred, no doubt attracted by the smell of the sweat released through the feet of the combatants during the duel. Some cats, mostly females and juveniles of both sexes, even approached toms engaged in a duel to investigate the odors they emitted from anal glands, but such intrusion rarely distracted a combatant more than momentarily. As a rule, other toms seemed happy to keep out of the way of two dueling toms.

While the females in the cattery sometimes spat and flashed claws at each other, I could count on two hands the number of times that I saw them actually duel in the manner of the toms. These duels, despite their rarity, proved that females also possess the full repertoire of instinctive actions associated with threat display, but that the threshold at which they are elicited is much higher than it is for toms. Although only one of the female duels that I witnessed involved the use of teeth and claws, all of them were very one-sided. The instigators seemed to be motivated by little

When two females start spitting and throwing blows at each other, most toms beat a retreat.

more than a rare overflow of aggressive drive which they worked off on any unfortunate female who happened to be in the vicinity, for I could detect no pattern in the choice of target. Kurokasan, whom I saw starting a fight on four occasions, chose a different adversary each time. Funnily enough, she was a full sister of Waru, and seemed to have the same aggressive streak. Having done nothing, at least to my eyes, to provoke such aggression, the attacked females seemed as surprised as they were scared, and showed no interest in actively responding to the threats of their attackers, merely crouching low and assuming a defensive posture until their tormentors desisted from further threats.

• • •

According to my observations, only about one tomcat duel in thirty escalated into physical clashes. Although this frequency may be much higher among free-ranging toms, the threat display of cats would appear to be a remarkably effective mechanism for the bloodless settlement of disputes. Despite the lack of bloodshed, the frequent noisy confrontations were sufficient to convince many naive observers that the cattery toms were an extraordinarily pugnacious crowd. I have to admit that I, too, with my attention always being drawn to the next duel, tended at first to overlook the measures which the toms were taking to avoid even nonviolent confrontations. While many duels were started through the deliberate instigation of one of the participants, just as many appeared to result from accidental encounters between rivals at various "blind spots" throughout the cattery; two toms would find themselves suddenly face-to-face at a passage entrance or when rounding a piece of furniture, and would feel obliged to assert themselves. I got the impression that most such confrontations could have been prevented had all the furniture and other structures been made of glass. Close observation of the toms' movements made me realize how well cats recognize each other at a distance of some yards, and how eager most of the toms were most of the time to avoid trouble.

I reckon that for every confrontation that occurred, ten were averted through the toms' keeping their eyes peeled and adjusting their movements to avoid crossing paths with rivals. Spotting a rival, quite often even one of inferior rank, snoozing ahead on one of the aerial walkways, a tom would make a U-turn or a detour along another walkway. If Panda was walking across the floor and saw Sha a few yards away heading in his direction, he would calmly but deliberately change his course, keeping Sha in his sights as he did so, while Sha for his part would appear to pretend not to notice Panda's move, almost as if out of consideration for the latter's pride. Walking around obstacles, toms frequently gave corners a wide berth in order to avoid possibly undesirable collisions with others. The more familiar I became with the rivalries between the toms, the more such behavior became apparent to my eyes.

If "traffic regulation" among cats is maintained through eyesight at short distances, there is another mechanism which would appear to play a similar role at longer range. I refer, of course, to the spraying habit of cats. Both toms and females spray urine, but the latter usually only infrequently. Since females of wild cat species tend to spray more rather than less often than males, the infrequency of this behavior in domestic females would appear to be one of the results (a most fortunate one!) of domestication. The fact that castration results in moderation and often total cessation of spraying in tomcats indicates that this behavior is connected with the output of the male sex hormone, testosterone, but the connection between spraying and sexual behavior is only slight. Toms will spray incessantly throughout the year, the frequency rising only slightly during the breeding season. Spraying directly backward at a vertical object comes most naturally to them, and so fence posts, tree trunks, walls, car tires, etc., tend to be their favorite targets; but they will squat and spray downward at any object or patch of ground that they consider particularly worthy of their signature.

Before spraying, toms also frequently rub an object with their chins, foreheads and sometimes tails, all parts of the body where epidermal scent glands are situated. The cattery toms frequently stood on their hind feet to rub their chins on a point higher than the one that they would subsequently spray, perhaps in order to separate the two types of mark and so increase overall effectiveness.

These spray and scent marks are thought, although concrete evidence is hard to come by, to provide information to other cats on the identity and movements of their owner and could thus be considered to play a part in the defense of territory, since cats no doubt take such information into account when deciding their further movements. However, cats do not react with fear to these marks, but investigate them at leisure and often spray over them before continuing on their way, suggesting that the marks are purely informational rather than olfactory "No Trespassing" signs. I mentioned earlier that tomcat territories in particular show a great deal of overlap, with different toms frequenting the same paths at different times. Spray and scent marks could be very important in the regulation of traffic in such communal and borderline locations.

It is not in the nature of instinctive behavior which has served the species well for millions of years to be scrapped as soon as it becomes inappropriate. Such was the case with the spraying of the cattery toms. At such close quarters, any information to be gleaned from each other's marks would have been redundant, and moreover, probably indecipherable in view of the number of toms present and spraying, but this situation made no difference to their behavior; driven by their instincts, they sprayed incessantly, here, there, and increasingly everywhere. Just as with any other aspects of their behavior, they showed individual differences in their spraying habits. Sha, a rather lazy tom, tended to save up for long and luxurious squirts, whereas Sabu, at the other extreme, was a tireless sprayer, often going through the motions even when he had only a drop or two, if that, to eject. Spraying locations also varied according to individuals, but many toms seemed to be willing to risk limbs if not life in order to mark certain spots. The uprights which joined the cylindrical horizontal girders, more than nine feet above the floor, to the roof supports were one such target. To get into a position to mark them, a tom would have to walk out along the three-inch-diameter horizontals and then perform a neat U-turn in order to bring his rear into line with the upright. I saw toms carry out this maneuver successfully many times a day, but in five years I must also have witnessed more than fifty occasions when a tom lost his footing and fell to the concrete floor after a vain struggle to right himself. I am sure that all of the toms were aware of the risks, but even those who had experienced a fall were not to be discouraged.

Over the long term, a tom's spraying frequency seemed to be pretty constant, as one would expect of an instinctive behavior pattern with its own propensity. However, certain situations called for particularly frequent spraying. Any new object introduced into the cattery, such as a cardboard box, a new piece of cast-off furniture, the trouser leg of a visitor, was immediately signed by a procession of toms. While pursuing an estrous female, far from forgetting to spray, they would be spurred to even greater efforts, often deliberately breaking their pursuit for a quick squirt here and there. After a duel, both victor and vanquished, once out of each other's vicinity, would invariably spray, suggesting that this behavior also has a kind of cathartic effect on strained nerves.

This was all rather interesting from the behavioral point of view, but I have to admit that there were times when I could quite happily have castrated every one of the toms if only to put an end to the habit. I normally gave vent to such feelings when my bare head or face was hit by a warm shower of urine delivered in total innocence by a tom spraying at some unrelated target along an aerial walkway above, but such occasions were the trigger rather than the root cause of my discontent. I had no great objections to being sprayed else-

Jerry succeeds in spraying this target. Many cats fell trying to get in position to do the same.

I am anointed by Otochan—an occupational hazard.

where on my person; anyone who did object strongly would not have lasted long in the role of keeper, since daily anointings were an occupational hazard. If anything, I was only too happy to be regarded as just another fixture, as worthy as any of their signature. Wellington boots gave ninety percent protection since most intentional sprays were performed at ground level. I was also impervious to the ammoniac odor that the toms' spraying imparted to the building, although this smell was enough to keep many a visitor, even professed cat lovers, from putting more than one foot inside the cattery.

Far more than the smell or any personal inconvenience, what I really objected to was the damage to the facilities and the inconvenience to the cats in general caused by the habit. Cat's urine possesses powerful corrosive properties, and its effects became increasingly visible with the passage of time and the increase in the tom population. Within eighteen months rust had so eaten spots in the wire mesh which screened the entrances and windows that I had to redo the lot with heavier-duty plastic-coated mesh. Within two years we began to see cracks of daylight appear at favorite spraying spots on the corrugated metal walls of the building, despite their paint coating, necessitating replacement and nailing of wooden panels inside at such locations to prevent the same thing happening again. Outside in the garden, the toms' urine proved to be just too powerful a fertilizer for the grass bordering the walls and other spraying spots. Despite frequent watering to dilute the effect of the urine, the grass there turned brown and then gave up the ghost.

As the number of toms grew, so did the number and variety of their targets; first the empty food bowls, then the tins full of dry cat food, and finally the two large water trays fell victim to the toms. We took to picking up the food bowls as soon as they had been licked clean, and to replacing the water three times a day and the dry cat food every other day, crushing up the tainted cat food for our hens, who ate it with relish and apparently no ill effects. The straw bedding used to line the cats' winter sleeping quarters was another spraying target and had to be replaced once or twice a week throughout winter. They had to do without straw in the warmer seasons! Toms even inadvertently sprayed each other on occasion. Looking back, the only good thing about the spraying was that I did not have to replace the sawdust in the tubs as often as I would have had to if the toms had used them at all regularly. In view of all the bother the infernal habit caused, I was heartily grateful that the females in general desisted from spraying, and often wished aloud that domestication had had the same effect on the males of the species. No doubt many owners of undoctored toms would echo such thoughts.

Although cats do not possess, in the context of a fight, any stereotyped aggression-inhibiting submissive gesture of the kind found in dogs, outside the context of a duel,

Mo tries kitten play to ingratiate himself with Aya.

they are no sloths when it comes to the use of gestures of defenselessness to inhibit aggression and make friendly contacts with other cats. All cat owners will be familiar with the way their pet approaches and rubs its chin and flanks against the human leg, soliciting attention in much the same way that a kitten does from its mother. An adult cat uses exactly the same tactics when approaching an acquaintance, mimicking juvenile behavior by offering its head to be licked in order to signal its friendly intentions. Both males and females appear to be equally sensitive to such solicitations, and will usually respond by licking even when their heart is not really in it. Frequently such approaches lead to mutual grooming sessions, after which both cats settle down together.

Other juvenile behavior may occasionally be employed. I once saw Mo, who had a very serious nature and almost never played, suddenly start to play with the tail of Aya, his long-standing sweetheart, whom he had been following around all morning with clearly romantic intentions. Aya, who was still occasionally nursing some kittens that she had given birth to three months beforehand, was not in heat at the time and had been all but ignoring Mo. It is interesting that Mo suddenly turned to "child play" at a moment when one of Aya's kittens was asking for her attention. It was almost as if, upon seeing Aya respond warmly to her kitten, Mo was struck by the idea that he could also gain entrance to her heart, where more direct demonstrations of his longing had failed, by acting like a kitten. Even though this tactic met with little success on this occasion, it is not uncommon for elements of juvenile behavior to be observed in the courtship not only of cats but also of many other mammals. Two dogs "in love" will often frolic and cavort with each other in much the same way as puppies at play.

Since juvenile behavior, symbolizing as it does harmlessness and vulnerability, is an effective means of inhibiting aggression, one could expect it to be employed not only for the ulterior aims of obtaining sexual favors, company, warmth or grooming but also as an end in itself by weak individuals wishing to convince stronger ones of their inferiority. Sure enough, in the cattery, we often observed young toms go out of their way to pester older toms for attention, rubbing against them, pushing noses at their faces, and so on. Relationships between young and senior toms were in general very relaxed, but occasionally we saw the same kind of thing going on between habitual foes.

The example that sticks most vividly in my memory is that of both Ee and Sha deliberately approaching Mo, whom they normally avoided, and getting him to groom them. The timing of these approaches was very significant, always occurring after mealtimes when, with his belly full, Mo was no doubt in the best of moods and in the grip of an urge, as all cats are after eating, to lick and lick, both himself and any other cat in licking distance; he thus obliged in a predictable way by licking down both cats, appearing almost not to notice who they were. This sight was to be seen almost every day for about two months, after which it occurred less and less frequently. Because, for Sha and Ee, this act was almost literally a case of putting their heads in the lion's mouth, so to speak, it surprised me greatly —even more so because it was not just a one-time happening. At the same time, however, it appeared almost unbearably familiar and understandable. After all, who but the proudest of us has not been guilty, at least once in our lives, of similar behavior? Ingratiating oneself with a feared superior at a time when he or she is likely to be most amenable to friendly approaches is a very common, if not very praiseworthy, human trait, and this is precisely what Sha and Ee appeared to be up to. Compared with the lengths to which we humans will sometimes go, Sha's and Ee's "boot-licking" was very mild, but I am nevertheless glad that the cats did not make a habit of it in general, and would like to think that the cause of its scarcity was as much the inborn pride and dignity of cats as fear of making such approaches.

As to the efficacy of Sha's and Ee's behavior, I was unable to note any significant change in the quality of their relationship with Mo, or with each other, for that matter. Although they had never been bitter rivals of his, Mo clashed with both of them occasionally, as they did with each other, before, during and after this period, and they appeared to avoid him as much as ever at other hours of the day. However, this should not be taken as evidence that Ee and Sha gained nothing from their efforts. For two short periods each day, they were able to feel safe in Mo's presence, and to receive a lick-down instead of a tongue-lashing. Who is to say that they did not gain something in terms of peace of mind, feelings of well-being or what you will, as a result? They certainly showed every sign of enjoying these encounters.

One more episode of a similar nature occurred early one morning in the spring of 1979 when, for some reason, Ee was temporarily infatuated with Hige, at that time still a young male. Hige was neither particularly encouraging nor discouraging Ee as the two sat among other cats sunning themselves on the window shelf of the far shed. He protested occasionally when Ee gripped the nape of his neck for any time, but did not try to escape when Ee released his grip. At this time Waru was already in decline as a disturbing influence in the place, and was not showing half the inclination to mix blows with Ee that he had two or three months previously. He was still, nevertheless, a force to be reckoned with, and so when he jumped up onto the shelf that morning and approached Ee and Hige, I half expected trouble.

It was with no small astonishment that I witnessed the strange sight that followed. Far from directing any threats at Ee, Waru approached a little closer after halting momentarily to take in the situation, turned around in front of Ee and then crouched down in what was a perfect female copulatory pose, his rump slightly raised in Ee's direction! He held this pose for two or three minutes, during which time he looked back at Ee occasionally and meowed in the same coaxing way that a female in heat sometimes does in order to draw the attention of a lukewarm suitor. There was absolutely no doubt in my mind that Waru was inviting Ee to mount him.

How are we to interpret this behavior? Could it be that Waru was overcome by a strange sexual urge, and his choice of Ee was coincidental? I somehow doubt it, as, for one thing, I never saw Waru behave in this way before or after this occasion and, for another, he knew Ee very well, which suggests that his choice of Ee out of all the cats in the building was no coincidence. All this indicates an ulterior motive behind Waru's behavior, and for want of alternatives I would like to hazard the following explanation. Waru was fed up with fighting Ee, a wish for less strained relations stirring inside him. Here at last he spotted an opportunity for expressing such a wish. Having accurately gauged Ee's mood, he assumed the posture of a sexually receptive female to ingratiate himself with his old enemy. I admit that this sounds very farfetched, but there are precedents. For example, among certain monkey species, weak males adopt female copulatory poses to inhibit aggression and express their inferiority and submission to superiors, who invariably respond nonaggressively by mounting. As it turned out, Ee appeared to be too wrapped up in Hige to take much notice of Waru, who eventually ceased his solicitations and settled down nearby.

Despite the rarity of the kinds of behavior described above, I think that they are

Waru invites his rival of old, Ee, to mount him.

very worthy of note. If my interpretation in terms of ingratiation is correct, they represent genuine efforts by the individuals involved to go beyond mere avoidance of hostilities, to attempt to get on friendlier terms with their enemies. While the cattery was a confined environment in which the occupants, unable to avoid each other all the time, even if they wished to, were under pressure to socialize to an unnatural extreme, I do not think that the kind of behavior shown by Ee and Sha toward Mo is necessarily unique to that situation alone. Leyhausen describes how free-ranging cats of both sexes, who are under no compulsion to socialize with each other, will often meet nightly on neutral territory to exchange greetings, engage in mutual grooming or just sit around close to each other.

Among the participants there might be cats which have been seen at other times fighting each other, but these meetings are apparently notable for the warm, nonviolent atmosphere that prevails. The cats attending these meetings would appear to be answering a very real need for social contact which must, however, remain suppressed at other times as the cats answer more urgent needs and instincts for survival.

I believe that the kinds of behavior that Sha, Ee and Waru showed were forms of expression of a desire for social contact, a desire to make peace with one's enemy and thus make life easier for oneself. I think that they also showed that cats are not completely lacking in a vocabulary of appeasement—they just use the vocabulary they possess very sparingly, and not within the context of duels.

A luxurious stretch from the boss.

8

"WHO'S THE BOSS?"

One of the questions most frequently asked by visitors to the cattery was, "Who's the boss cat here?" When one thinks about it, what could be a more natural question from a human being, an animal who so invariably lives in stratified societies peppered at all levels with leaders? It is only an individual with specialist knowledge or a keen imagination who will observe a group of animals and not immediately assume the existence of a boss. If he is observing a group of naturally social animals, preferably in their natural environment, he will in many but not in all cases be able to distinguish an individual who fits the bill, but what we had in the cattery was far from a natural environment, and the cat, moreover, is not an animal naturally given to forming large social groups. What use would an animal that has for millions of years led a solitary existence have for a boss?

As any cat owner knows, cats are their own bosses. A cat will decide for itself when to get up, when to sleep, when it wants to be fed or stroked or left alone. When it wanders outside, it wanders alone. Even the cattery cats, although they came together to sleep, to eat, to fight, to play, to mate, were strict loners when it came merely to wandering around the building; to go for walks together is for cats a totally foreign concept. In their relationships with each other, there are strong cats and weak cats, but a strong cat shows as little inclination to lead or to use its power to secure privilege as a weak cat does to be led or to knuckle under. Cats are born individualists or, to use a political metaphor, anarchists, and their social relationships tend to be two-dimensional and egalitarian rather than three-dimensional, hierarchical and authoritarian. For these reasons it was always with some hesitation, when asked, "Who's the boss?" that I pointed out Mo.

After he had overcome Waru's challenge, Mo was indisputably the number one tomcat in the place, and as such sat at the top of the hierarchy that had evolved among the toms. This hierarchy was, however, of a very loose nature. Immediately below Mo were a number of toms such as Ee, Waru, Wanchan, Sha, Sabu, Nibu and Panda, all of whom seemed to owe their status as much to their age and the fact that they were the first mature toms to inhabit the cattery as to their fighting ability. These

In the cattery, the toms tended to fight mostly with those of their own rank.

toms fought mostly among themselves and with a few slightly younger toms, more on the basis of personal enmities than on any firm ranking. Thus, for example, Sha fought with Panda, Wanchan with Nibu, and Ee with Waru and Panda, but Panda, Nibu and Sabu, who had grown up together, never fought with each other — likewise Ee and Wanchan. However, it was by no means the rule that toms who had grown up together never became enemies. As with humans, former friends, even brothers, often came to cross swords the most fiercely. Such was the case with Jerry and Otochan, two of the litter of five kittens whom we had hand-reared. Needless to say, it was with some-

Toco and Chi with their first litters, while "Granny" looks on.

Mothering five orphaned kittens was a full-time job. Their first efforts at eating solid food were messy, to say the least.

Within a short time, however, Otochantachi had become quite proficient eaters.

Playing "lighthouse."

The kittens thrived on chaos.

Toco in the woods.

Now they could exchange morning greetings with the other Kingdom dwellers.

Exploring the countryside with Otochantachi was a memorable experience. Here Hi tackles tree climbing.

"You smell of grass." "And you smell of fish."

Otochan with a nosy Shetland pony.

Snoozing contentedly in the cattery.

Toco after sparrows.

They were able to spread themselves on innumerable perches that caught the sun at different times of day.

Harpo sunning himself in one of the garden's trees.

I firmly believe that a cat should be able to roam as he pleases—a free cat is a happy cat.

Five apprehensive but resolute cats on the seashore.

The first sign that the mating season had begun would be a crowd of toms excited by a "hot" smell.

Chase that lady!

The estrous female, Mu, is third from the top, the rest being hopeful toms.

"Keep away from me, you randy lout!"

As soon as the female begins to howl, all the other toms in the vicinity begin to think about escape.

An unexpected maternal response. Hearing the distress call of her seven-month-old kitten, Cherry, Pe suddenly responded to him, forgetting the suitors on her trail.

Fumi trying to entice her last, halfhearted suitor to make love.

The estrous female is the black and white cat in the forefront. Almost all the rest are her suitors, but she won't remain so popular for long.

Female orgasm—a vital prerequisite for ovulation?

what mixed feelings that we watched the battles between these two, who had in their kittenhood shared so many adventures together under our care. As another example, Pierrot, a young tom of weak character himself, was one of the most frequent persecutors of his own brother Kagetora, the most notable pariah in the history of the cattery.

The lower-ranking toms fought less frequently than the "seniors" mentioned above; here again there was no evidence of a firm "pecking order," confrontations between these "juniors" being determined mainly on the basis of personal enmities. Juniors rarely challenged seniors even as they grew older, and seniors left juniors alone for the most part. While the seniors treated Mo with the utmost respect, most of the younger, low-ranking toms displayed almost no fear of him. I once even witnessed the comical sight of Jerry, when already a sexually mature tom who also had a fair amount of fighting experience, mount Mo one spring morning in a fit of sexual excitement. Mo just crouched quietly until Jerry released him, and sought no redress for this impudence. Although Mo frequently instigated duels with other seniors, in particular Waru, who grew to live in mortal fear of him, he showed absolutely no interest in impressing his superiority on the juniors.

I should point out here that my division of the toms into seniors and juniors is pretty arbitrary and refers to two archetypes rather than real divisions. There was no sharp dividing line between the two, many toms, such as Hige, Osato, Marshmallow, Kenbo, Chobikuro and So, falling between these two in terms of rank.

Because most toms fought, through design or accident, with adversaries of much the same strength, one-sided confrontations were infrequent. Seniors would occasionally work off accumulated aggressive drive by bullying juniors, but they normally had their hands full dealing with each other. By far the greatest number of one-sided confrontations occurred between low-ranking juniors and those unfortunate toms occupying the slot, at the very bottom of the hierarchy, reserved for pariahs. At any one time there were always one or two young toms, invariably individuals crossing the bridge from adolescence into adulthood, who became the targets of repeated attacks from other young toms, many of whom were pretty weak characters who had gone through the same treatment themselves. It was as if pariahs provided a channel for the aggression of those cats without the strength of character to face up to their equals, although very occasionally senior toms and also females would avail themselves of the opportunity a pariah offered for some easy "sport"—for "sport" seemed to be the ideal word to describe these encounters from the point of view of the aggressor. At times I could not help feeling that the latter were working off not only pent-up aggression but also a pent-up hunting drive, since the pariahs, with their habit of fleeing before eventually being forced to face their foe, were ideal prey substitutes.

• • •

Until Kagetora's emergence as a long-term pariah, I was convinced that "pariah-hood" was just a temporary state, a somewhat extreme form of a rite of passage into adulthood which the majority of adolescent toms accomplished relatively unscathed. All of a sudden one day, we would notice a young tom cowering in a corner or up high on top of a pillar, eyes dilated and ears flattened, growling and spitting at any cat, male or female, who came near him, and we would realize that we had another pariah on our hands. He would usually have to be coaxed down to eat at mealtimes, when he would wolf down three or four huge mouthfuls and then scramble back to his retreat while the other cats were still eating. If he was unlucky, he would draw the attention of a spiteful persecutor who would give chase, nipping and slashing at his hindquarters. Most pariahs bore scars, old and new, on their rumps and hind legs.

Of course we felt sorry for such individuals, but experience taught us that almost all of them, once over their initial terror, soon learned, by keeping quiet unless directly threatened, not to draw unnecessary attention to themselves. A pariah would wait for times of the day when most cats were sleeping to come out of the shadows, make use of a sawdust tub and fill up on cat food and water. He would still be apt to flee at the slightest provocation, but would soon become quite adept at the art of escaping, and would not be as rattled by attack as he was formerly. Little by little his persecutors would leave him alone, and soon he would be moving around with greater freedom. When he began to join in meals of his own accord, we could breathe a sigh of relief, reasonably confident that the worst of his ordeal was over.

We noticed that those most liable to go through this stage of persecution were, ironically, youngsters whose confidence, friendliness, and general *joie de vivre* set them apart from their contemporaries. Such youngsters always appeared to be the most eager to put their growing strength and awakening aggressive drive to the test, their impudence eventually bringing about their downfall. Their confidence would desert them entirely after they had received their first sound thrashing at the hands of a superior foe.

Kagetora was just such a youngster. He received his name, which translates as "Shadow Tiger," at the age of three months, the day after I had watched a historical drama on television about the life of a famous Japanese samurai of the same name. He was the first blue tabby to be born in the Kingdom, and was a big, handsome and outgoing kitten, by anyone's estimate the pick of that year's crop, despite his tail, which was about half the length of any average British cat's. The name seemed perfectly to fit his nature and his coloring of dark gray stripes on a lighter gray background, and already at the age of ten months he showed all the signs of living up to it, when I first saw him instigate a duel with a tom twice his age, forcing the latter to adopt the defensive attitude of a distinctly inferior adversary.

It was only a month later, however, that

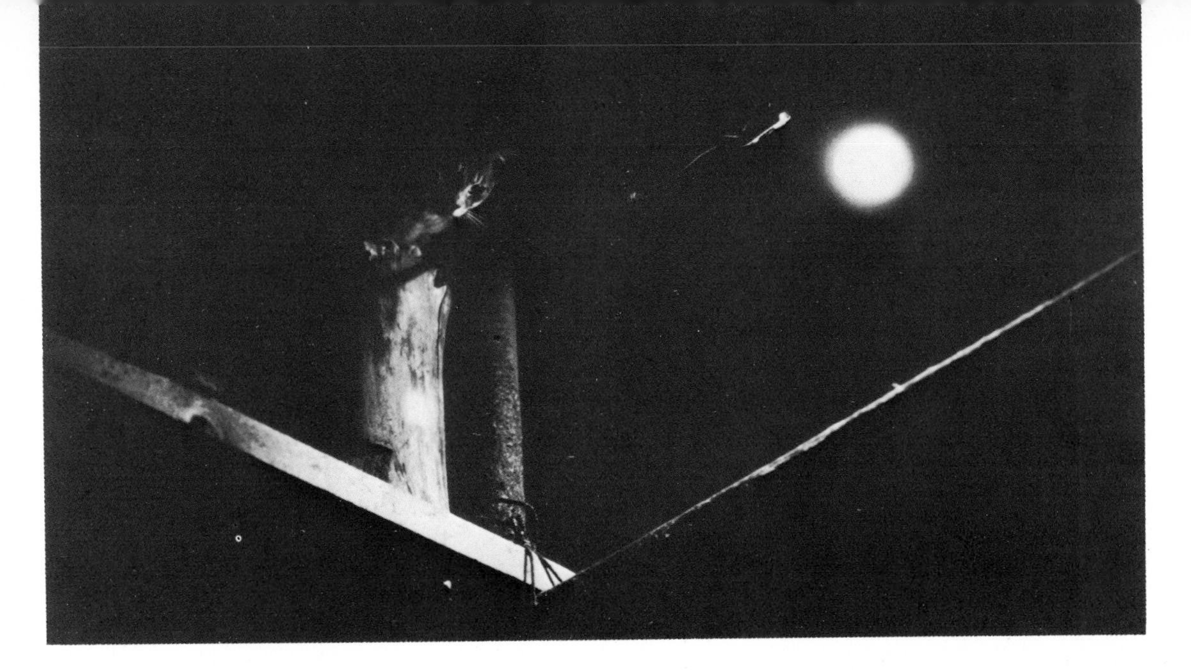

Kagetora, safe but lonely, remained on top of his pillar throughout the night.

Kagetora appeared to have received his comeuppance. He invariably led the horde of cats who, upon hearing me enter the cattery early each morning, came rushing to beg for their treat of *niboshi*, but one morning he did not appear. I discovered him as I made the rounds of the tubs to replace sawdust, atop one of the pillars set in two rows in the far shed. When he refused to descend in answer to my coaxing, I immediately suspected that he had become a marked cat, that his impudence had finally attracted the attention of a superior adversary or adversaries. The events of that morning proved my suspicions to be correct, for in the space of two hours I saw three toms, all of them juniors, and one female try to get at him. Where pariahs were concerned, news seemed to travel fast in the cattery, and by the end of the day, the number of Kagetora's persecutors had doubled. Luckily for him, cats possess little conception of "group action," and thus they do not make concerted attacks on a pariah, but the intermittent attacks of an assortment of single cats put him in a state of permanent terror. On top of the pillar, he was in a pretty impregnable position and managed to force his assailants back down by raining paw blows on them, but it was clear that he had no intention of moving from this retreat; he was in exactly the same place, with the same look of fear on his face, the following morning.

I thought that, with the passage of time, Kagetora would get over this crisis as had the other pariahs, but he proved to be an extreme case. He became as adept as any of his kind at escaping and defending himself from his retreat, but just as he was showing signs of relaxing and gaining confidence, another thrashing would put him back at square one. Because I did not come across any feces at the base of his pillar, I knew that he was descending of his own accord occasionally, if only to use a lava-

Kagetora panting in 80°F.

tory. Figuring he just needed more time, I did what I could to make him a little more comfortable, fitting a platform of about one and a half square feet to the top of his pillar so that he could at least stretch out a little. He responded simply by transferring his perch to the next pillar. Certain that I knew what was best for him, I proceeded to fit similar platforms to every pillar. He now had little choice but to avail himself of one of them. He chose the furthest one from the entrance to the far shed, and began to use the one next to it as a lavatory. So much for my bright ideas! He now had even less excuse for venturing down from his pillar-top world. It was midsummer, and while daytime temperatures outside hovered very pleasantly around 65°F, the temperature directly beneath the rafters of those two metal sheds often hit 80°F. On the hottest days most of the other cats would snooze out in the garden or on cool patches of floor, but Kagetora would pant through the day (incidentally the only cat that I have ever seen pant) up on his platform, only daring to rest on the aerial walkway a yard below when there were no other cats around, and when he could bear the heat no longer.

His hind legs were crisscrossed with scars.

Autumn brought Kagetora relief from the heat, but winter was soon upon us, and he now faced the opposite problem of fighting the cold. I knew how cold Hokkaido winter nights could be, and the thought of Kagetora passing them alone up on his platform, while almost all the other cats snuggled up to each other in warm quarters, worried and saddened me, but I delayed taking any action in the hope that the cold would do the trick and force him to make a move where the heat and all else had failed. Here again, however, Kagetora proved more stubborn than brave. When he showed no signs of abandoning his platform even as New Year approached, my pity got the better of me and I rigged up a straw-lined box next to his platform. At the same time, I also gave up coaxing him down to ground level to eat with the other cats, feeding him instead up on the aerial walkway near his box. He finally had me beaten.

In hindsight, my efforts to improve his lot had in all likelihood hindered rather than helped his "rehabilitation." Had I forced myself to be totally cold and objective, and just left him to his own ends, he would either have had to pluck up courage and face down his tormentors, or he would have perished. I had found it impossible to be so coldhearted as to push such a choice on him, but had been too stubborn and dumb to realize then that the only responsible and humane alternative was to remove him from the building. For over six months I had been telling myself that time would do the trick, that the other cats would forget Kagetora, and that he would forget his fear. During that time every small sign of improvement in his behavior had brought me hope, and every reversion despair and anger. At one and the same time I had been filled with contempt for his cowardice and admiration for the tenacity of his will to survive. However, with the realization that his condition was in all likelihood chronic, that in spite of or because of my efforts, all he was ever going to do was just survive, I was left only with a feeling of disgust for myself for having put him through such unnecessary torment. When a New Year visitor to the Kingdom expressed a wish for a male kitten, I told her about Kagetora, and she immediately offered to adopt him. No one could have been happier than I at such an outcome, except perhaps Kagetora, once he had settled into his new home. His ordeal was finally over.

Lacking external outlets for the aggressive drives of its members, any confined community is a potential pariah-creating situation. The unfortunate individuals selected as pariahs for reasons of their weak physiques or certain character traits (or often, in human communities, because skin color, religious belief, social customs or political persuasion differ from the majority) play the role of safety valves for the release of aggression which, if unfocused, might tear the community apart. Evidence from the cattery, other animal communities and our own human societies shows that those most likely to focus their aggression on pariahs are invariably pretty weak or underprivileged characters themselves, whose low social status provides them with no other outlets. The more crowded and confined a community is, the more the apparent need for pariahs. I could thus take some consolation from the fact that, apart from Kagetora, no other pariahs of permanent status emerged. The cattery would appear to have been spacious enough to enable its occupants to do without the safety valve that a pariah provides.

After this grim picture of life for a pariah cat, my earlier assertion that the social structure of cat "societies" is basically two-dimensional and egalitarian would appear to have a false ring about it. There was undoubtedly a great difference in the quality of life led by Kagetora and that led by, for example, Mo; however, I stick to my assertion. For one thing, pariahs were the unfortunate offspring of an unnatural, confined community, and moreover they were the only individuals whose low social status affected other aspects of their lives, such as feeding and mating, to any great extent.

The only criterion upon which I based my arbitrary division of toms into seniors and juniors was who fought with whom. The very loose hierarchy detectable as a result had few implications for most aspects of life in the cattery. As I mentioned in the chapters on the cats' love life, Mo and the seniors rarely used their muscle to monopolize estrous females. Where resting places and feeding were concerned, we observed a similar adherence by most of the cats most of the time to two simple principles: "first come, first served" and "live and let live." Leyhausen found, in his own artificial cat communities, that unless the cats were very hungry, the top-ranking tom was left to eat on his own and that weaker individuals left the food bowls earlier. Except where the few pariahs were concerned, such was not the case in the cattery.

Many strangers, watching the way our cats attacked their twice-daily meal of boiled fish and rice, immediately got the impression that we were starving them, such was their display of voracity. We sometimes had to point out to such people the tins of dry cat food which were always available to the cats, and which they ate with enough relish outside mealtimes. There was no doubt, however, that the cats considered their fish meals to be the main events of each day. They tended to wolf down this food, not because they were particularly hungry, but simply because their neighbor was doing likewise. A cat who stuck to the "ideal" feline method of eating delicately and leisurely would have been at a distinct disadvantage. We fed the cats at the same times each day, and they knew these times exactly, beginning to rouse themselves and gather at the entrance to the building from about twenty minutes beforehand. (They demonstrated the accuracy of their internal clocks in their reactions on occasions when circumstances forced us to delay their mealtimes. Any delay of over half an hour had a distinctly adverse effect on their appetites. Far from displaying increased enthusiasm, they appeared almost subdued and disgruntled and took far longer than usual over eating.)

When we entered the building, they would rush in unison toward us and meow and mill urgently around our feet, those cats with tails raising them in elegant, tremulous curves in their excitement. As we doled out the food, about four or five cats would gather at each bowl. All would fold back their ears and whiskers as far as possible, appearing to dislike the sensation of ears and whiskers touching those of neighbors. While a few famous gluttons, like Ee, Waru and the female with the upturned eyeballs called Uemi ("Eyes-up"), kept their faces firmly in the bowls as they gulped mouthful after mouthful, most cats reached forward to fill their mouths and then retreated slightly to chew and swallow while others pushed to the front to grab another mouthful. Some of the cats developed the habit of using their claws to hook chunks of fish from bowls rather than go in with their heads. Even more numerous were those who used their paws to pull a dish toward them. With two or three individuals up to this trick at the same time, a bowl would be pulled this way and that, but such tugs-of-war never escalated into angry exchanges.

After as little as two minutes, the gratifying sound of tens of feline jaws smacking would die down, as one cat after another left the "dinner table" for the water trays or favorite spots in which to lick down themselves and each other. Before doing so, there were inevitably a few toms who would take care to anoint empty bowls with their urine. A few other cats, the inveterate plate-lickers, would linger a little longer to polish off every last grain of rice to be found in or around the bowls, and then the meal would be well and truly over—all in the space of five minutes from the time that we had entered the place.

Mo would get very ratty just before a meal, particularly with Waru and other rivals, but he never hogged a dish to himself.

Unlike Leyhausen's cats, at no time did we see high-ranking toms attempt to hog bowls to themselves, or low-ranking toms and females giving way to superiors. As with the mating with an estrous female, mealtimes were also a free-for-all, with the fastest eaters, not those highest in rank, profiting at the expense of more delicate feeders. This, to my mind, was a far more egalitarian form of food-sharing than one based on rank.

There was, however, one cat who habitually monopolized a bowl, and his case was an eloquent illustration of the kind of attitude which the cats adopted toward their lives in the cattery. His name was Petoru, and he had been found with his sister abandoned in a cardboard box outside the Kingdom gate. Being already mature, these two cats had had trouble settling down in the cattery, and Petoru in particular appeared to object strongly to communal eating. For the first few weeks he ate nothing but dry cat food, but as he gained in confidence, he came to wait near the end of the row of bowls, and as soon as the bowl that he had earmarked had been filled with fish, he would sit over it and begin to eat at a leisurely pace. At the same time he would make it clear, by hissing and flashing his claws at any cat apart from his sister who approached, that he would not tolerate any sharing. In a short time, all the cats knew of this quirk of Petoru's and steered clear of him when he was eating. I was always impressed by the cats' tolerance despite their obvious desire to share Petoru's bowl, which was invariably still half full when all the others had been emptied. Perhaps because he was a recent immigrant and anyway of a very gentle nature, Petoru showed little inclination to get involved in duels, holding his own ably enough on the few occasions he was challenged, but rarely instigating a fight himself. The cats' avoidance of him at mealtimes was thus, in my opinion, born not so much out of fear of Petoru himself, but rather out of an inherent respect and fear of any cat who is clearly occupying a certain spot and insisting on its right to be left alone there. On occasions when he was too late to secure a bowl to himself, Petoru in a way showed the same respect for this first-come, first-served principle by desisting from any attempt to chase other cats away from a bowl, and eating dry cat food instead.

The same principle operated with respect to resting places. With the exception of pariahs, any cat of any rank could settle on an unoccupied spot without much fear of being dislodged. A tom would sometimes vacate a spot of his own accord upon seeing an approaching superior with whom he was on bad terms. Failing to budge did not, however, in most cases provoke the aggression of the superior who would either pass by unconcerned or change his own direc-

Petoru warns So to keep away from his dish. The cats showed surprising tolerance in leaving him to eat alone.

tion. Even if a cat seemed to be determined to dislodge a lower-ranking rival from a particular spot, there was no guarantee that he would succeed. I once observed Mo approach Sha with obvious aggressive intent when the latter was basking on a sunny patch of floor. I don't know whether Mo was spoiling for a fight or merely wanted Sha to move, but Sha refused to oblige. He just continued to lie there on his side, pretending not to hear Mo's low growls and averting his eyes from Mo's threatening gaze, the slight twitching of the tip of his tail the only indication that he was only too well aware of Mo standing over him. Sha's whole attitude seemed to be expressing his unwillingness both to mix words with Mo and to get up and move. If one was to put words in their mouths, Mo would have been saying to Sha, "How dare you just lie there, you pipsqueak? Don't you know who I am?" and Sha would have been replying, "Come on now, old chap, don't be such a cad. I've got no argument with you. If you leave me alone, we'll save each other such a lot of bother."

Whatever Sha's message to Mo was, it seemed to do the trick. Mo continued to glare and grumble at Sha a little, but he was already slowly withdrawing. Sha would never have got away with such nonchalant behavior had Mo confronted him when he was up and moving around the cattery. He was able to do so because by (albeit unwritten) common feline law, he had much more right, as established occupant, to be there than Mo. This law was by and large upheld by all of the cats, including Mo, and incidents like the one just described were not the rule.

Personal character could possibly play a part in determining the degree to which a cat uses his rank to his advantage against inferiors. Had Waru managed to hold on to the number one spot, the chances are that he would have been a much more despotic top cat than Mo, who was at most times an exemplary gentleman. However, the extent to which even the most despotic cat could exert authority would be severely limited by the unwillingness, you could even say inability, of cats in general to knuckle under and accept such authority.

I found myself often comparing our cats' fairness to each other (again with the exception of their treatment of pariahs) favorably with our dogs. The dogs' social structure was far more hierarchical, with the dogs at the top frequently exercising authority to advantage, if we let them. It was only our presence at mealtimes which prevented the stronger dogs from helping themselves to the bowls of the weaker dogs, who would otherwise leave their food almost untouched at a glance from a superior. The top group of big senior dogs occupied the area immediately outside the main house, which was considered by all the dogs as prime territory. A larger group of weaker dogs occupied the stables, venturing as far as the main house only when in the presence of people, and skedaddling back to the stables as soon as those people disappeared into the house. One borderline pariah lived behind the stables, and would eat only if her bowl was placed a few yards apart from those of the other dogs. Her location offered the best view of the gate a couple of hundred yards away, and I got the impression that she was permitted by the other dogs to lead a reasonably peaceful, albeit restricted, existence only because she was useful to them in the role of watchdog, always being the first to bark and warn of the arrival of strangers. One more dog, a castrated male, was a genuine pariah, and until we moved him to separate quarters to alleviate his misery, survived only by spending almost all of every day in a quite expansive hole that he had dug under Mutsu-san's study. He would only come out to eat and relieve himself immediately outside his hole when the other dogs were eating in the stables, or in the dead of night. Occasionally the dogs got hold of him when he was careless, and it was only our quick intervention with swinging baseball bats that saved him from certain death. In general, the lower-ranking dogs had far more problems dealing with their superiors than the lower-ranking cats with theirs.

As I have already explained, the cattery was essentially a nonterritorial environment, with almost all the cats using the whole area available to them with relative freedom. Their practice of a "live and let live" philosophy was very much a result of their consciousness of the cattery as neutral territory. A free-ranging cat, tom or female, could not be expected to show the same degree of tolerance to a strange cat it found snoozing in its territory. It was because the cattery cats regarded their communal quarters as neutral territory that we were able to introduce strange adults without having to worry for their safety. Such strangers were almost always understandably petrified by the experience of suddenly finding themselves in such a strange place, the center of attention of a horde of unknown conspecifics, but they were never attacked.

We could never take the same risk with a strange adult dog, since our pack had very set ideas about the presence of strangers within their territory. On two occasions they showed no mercy to unfortunate intruders who, drawn by the scent of our bitches in heat, had broken in through the fence, and found themselves suddenly confronted by the whole angry pack. For this reason, it was impossible for us to introduce strange adult dogs into the Kingdom, and if we could not rapidly find new homes for the various adults abandoned outside the Kingdom gate, we had no choice but to put them to sleep.

Even if the cattery was as a whole neutral ground for its residents, the first-come, first-served principle by which they lived could be regarded very much as an expression of territorial instincts in an environment which did not permit the establishment of permanent, tangible territories. Just as a cat, when well inside its territory, will often prevail over a trespassing neighbor who is of higher rank in absolute terms, a cat occupying a certain spot in the cattery could, if disturbed, be more than a match for a superior. In other words, the same forces of mutual repulsion that are the basis for

Mo with sister May.

territory formation in a natural environment found expression in the cattery in a form of "mobile territoriality," with each cat appearing to carry territory around with it, this territory being most evident when a cat settled down in a certain spot, and recognizable by the courtesy shown to such a cat by others.

Territory size is very much affected by population density. While I would object to a stranger sitting too close to me on a virtually empty train, for the dubious privilege of riding a rush-hour train into Tokyo, I, along with everyone else, have to suppress my territorial instincts entirely as we pack a car to what the railway authorities blithely refer to as three hundred percent capacity, a condition which would shock sardines themselves, let alone the most stalwart of commuters on the most crowded of British trains. In a similar way, our cats would take little notice of each other as they crowded onto the east-facing window shelf in the early hours of the morning to take in the warmth of the rising sun, but at other hours, when the shelf was pretty bare of life, the cats using it would space themselves out or, when they approached each other, would do so with far more care.

"First-come, first-served" and "live and let live" are hardly the kind of mottoes that a "boss" would be expected to live by. Whether the boss is a despot or a democratically elected leader, the exercise of authority, performance of duties and access to privileges are, to a greater or lesser extent, the features of his or her life. We have already seen how limited Mo's authority was, and how disinclined he appeared anyway to use his muscle to monopolize food bowls, resting places and females of his choice. He showed an equal disinterest in taking on any of the duties that befall many a leader of groups of social animals. He showed as little interest as any others in flushing out a strange newcomer to the cattery. And although defense of a social group is normally a task taken on by its strongest members, Mo would be among the first cats to flee to the rafters if I brought a dog into the building.

Another duty often performed by a boss is the maintenance of peace within the group, protecting weaker members by breaking up squabbles between his inferiors. There were times when Mo appeared to be doing exactly this kind of work. I remember one occasion when Sha provoked a quarrel with his pet hate, Panda. Wanchan, passing by at the time, decided to butt in, focusing

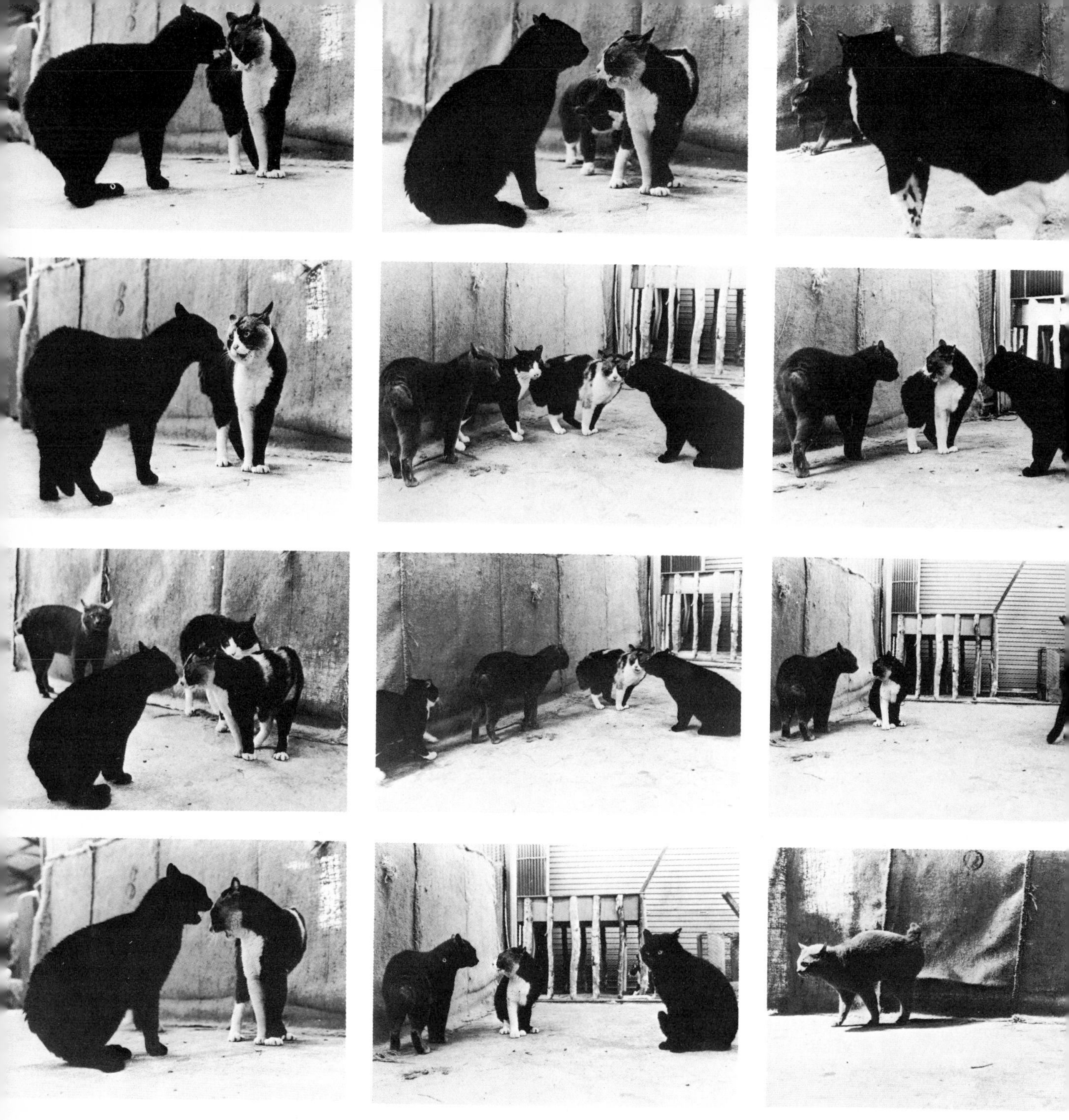

Mo intervenes in a tussle between Sha, Panda and Wanchan. He was concerned only with keeping rivals in check, not lording it over the whole population.

his attention more on Sha than on Panda. Drawn by the sight and sound of all this name-calling, Mo approached threateningly on stiffened legs. Wanchan, closest to Mo both in location and rank, was the first to back out of the hostilities, and then Sha also signaled his lack of enthusiasm for tangling with Mo by sitting down, turning his head away, and with great trepidation making a quiet exit. Mo glared at the one remaining combatant, Panda, for a few seconds, then sprayed and stalked off.

Maybe because altercations involving four cats were so rare in the cattery, and because this one was an almost too-perfect demonstration of the ranking order existing between the participants, it stuck in my memory. Mo finished up looking very much like a king who had used his rank to bring about the peaceful settlement of a dispute among his underlings; but the fact that he intervened in such disputes only occasionally and completely ignored those occurring between toms much lower down the ladder, suggests that he was far more concerned with impressing his superiority on his immediate inferiors than on creating peace in the cattery.

With few privileges and virtually no duties, Mo hardly filled the popular image of a boss. Although he was the top cat in terms of fighting ability, he clearly was not, nor did he show any inclination to be, lord of all he surveyed. There was no doubt that he sat at the top of a hierarchy, but the influence of his hierarchy upon life in the cattery was far from pervasive. Pariahs were the only cats whose lives were radically affected by their status, the remainder of the cats appearing to enjoy roughly equal rights, irrespective of rank. Mo, in his disinterest in consolidating and extending his authority over the social structure, seemed to symbolize its egalitarianism. Far from being a boss, he was merely the strongest among equals.

The word "society" is an equally inappropriate term to describe what we had in the cattery, which was nothing more than an unnatural conglomeration of cats. Despite the existence of a hierarchy among the toms and a host of more horizontal relationships, both hostile and friendly, the population was bound together as a whole not by any common needs, purposes or sense of group identity, but simply by the walls of the building. This is not to say, however, that the cats, given a choice, would all have preferred to lead solitary existences. Studies of more natural populations show that cats will very often choose to live together in small groups even though they are under no compulsion to do so. The important feature of these groups is that females, moreover blood-related females, would appear to be at their core. Such groups were almost impossible to perceive in the everyday life of the cattery, such was the general atmosphere of amiability among all the females, but that amiability, plus my observations on kitten-rearing among the females, appeared to support the notion that society, if the word can be applied to cats, is to be found in the relationships between the females rather than the males of the species.

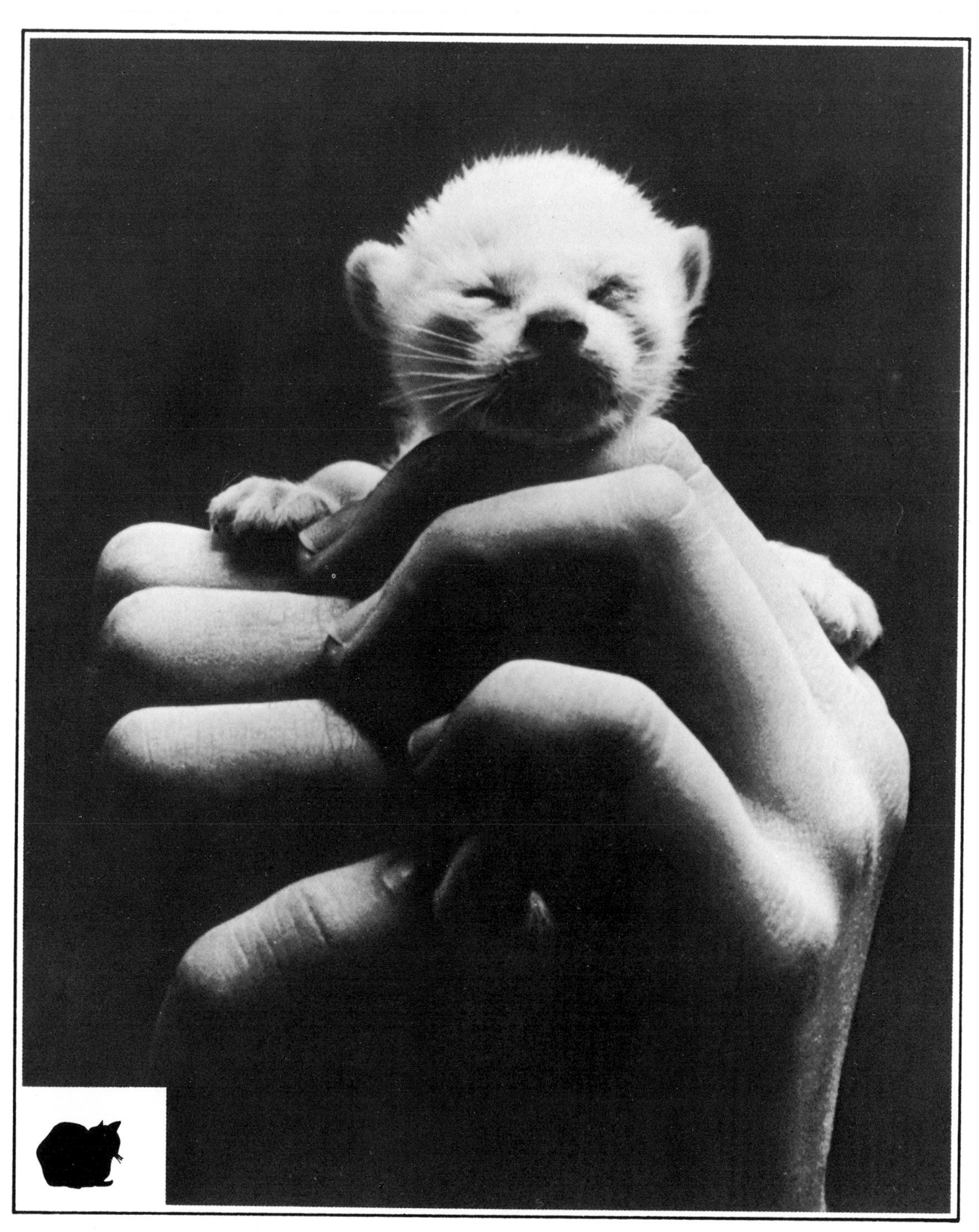

I never tired of the sight of new life in the cattery.

9

BIRTH WATCHING

It was shortly before two in the morning on June 5, 1980, when Pe's contractions finally began in earnest. "And about ruddy time, too!" I remember thinking, despite my relief that my calculations and observations had proved correct and my vigil had not been in vain. At sixty-three days after her first observed mating, Pe was dead on time. At sixty days I had put her into the small "delivery room" that was located in one corner of the far shed. Since she had given birth in the same place the previous year, she had recognized where she was and what she was there for, and had settled down quickly. Her general restlessness and lack of appetite on day sixty-two (up until then she had been eating enormously) plus a sudden extra tautness to her already swollen breasts informed me of the imminence of birth.

I never tired of the sight of new life being born, and was always interested to find out what strange new colors and patterns decorated newborn kittens. At the same time, I was generally busy with work, both in the cattery and outside, from about six in the morning until dinnertime each day, and am anyway by nature diurnal in the extreme. In other words, the prospect of night-long vigils never excited me, and since I was by that time no stranger to the sight of kitten births, I went out of my way to be present or close at hand only for first-time deliveries by nervous young females, just in case my assistance was required; proven veterans like Pe were left to their own devices unless I happened to find them giving birth at a civilized hour of the day.

However, I had reasons for making an exception of Pe on this occasion. She was not only an ideal cat for observation and photography, being an excellent mother, and welcoming rather than resenting human company even at childbirth, but she was also the last female due to give birth that year. This meant that I would be able to give her and her kittens more room than those before her, and observe and record her child-rearing under more ideal circumstances. Since I had selected her for special attention, I could not very well excuse myself from attending her delivery, and so after dinner that evening I had returned to the cattery equipped with a thermos of coffee, a book and a few tidbits for Pe.

I had been certain that she would not keep me waiting long, but as invariably happens when one makes preparations for a vigil, that is what one gets. It was almost as if Pe, seeing me come prepared for a night's stay, had decided that there was no need to hurry. She spent the next few hours alternating between the nest I had prepared for her, where she snoozed in fits, and my lap, where she purred solidly. It was past one o'clock, long after most of the cats outside the room had ceased scampering, meowing and occasionally caterwauling and had settled down to sleep, just around the time when I, too, was having increasing trouble keeping my eyes open, that Pe began to show signs of real restlessness, turning frequently in her nest and licking her vulva energetically.

The first kitten appeared at one forty-five, its little pink nose clearly visible within the balloonlike transparent amniotic sac which protruded further out of Pe's vulva with every contraction. This sac is more often than not ruptured before the kitten appears, but on this occasion it burst as one more contraction finally forced the kitten all the way out, wet and slimy and wrapped in membranes. Pe wasted no time in propping herself up on her forelegs to reach around and put her tongue to work, paying as much attention to her rear end as she did to the kitten, which wriggled healthily in response. It was soon free of membranes, and Pe began to chew at the umbilical cord, breaking it off about an inch from its attachment to the kitten's belly. As she continued to lick busily and occasionally tug at the cord protruding from her vulva, the afterbirth appeared, and then almost as quickly disappeared into Pe's mouth.

A mother severs the umbilical cord of her first kitten. Newborn kittens were a powerful prey stimulus to young cats and could not be left to the mercies of the cattery.

Some veterinary textbooks recommend preventing mother cats from eating afterbirths because they would appear to have a laxative effect on the mother's bowels. I, too, had noticed that some females showed mild diarrhea for a day or two after giving birth, but not of a degree to cause any worry. The laxative effect may even be of adaptive advantage, since expelling hard stools during the first few days after childbirth might be painful. Interestingly, it is not only the mothers of carnivore species who eat their afterbirths. Iwase-san, whose veterinary specialty was horses, told me that mares will sometimes eat portions of an afterbirth, and that many old horse hands, at least in Japan, are glad to let their mares do so, believing that the afterbirth contains substances which stimulate the flow of milk. Whether there are scientific grounds for such a belief I don't know, but I, too, was content to let the cats follow their natural instincts in this matter, instincts which, in view of the reproductive success of the species, would appear to have served them very well.

Pe's first kitten was an orange and white male. The second, to my delight a blue-cream tortoiseshell, was born half an hour later, and was followed at roughly twenty-minute intervals by another orange and white, this one a female, a tortie of the same coloring as Pe, and finally a black and white male, making a total of five. Both the tortoiseshells were females—it was too much, I suppose, to expect Pe to produce a tortoiseshell male on top of delivering with such thankful regularity and speed, but I was always on the lookout for one of these rare creatures. For two hours after the birth of the first kitten, Pe was the picture of busyness, resting her tongue from its tireless activity only when her contractions were pushing another kitten out into the world. She purred occasionally even while she worked, but now, with five dry, healthy kittens snuggled up to her chest, she gave herself over to a continuous deep rumble of a purr which only increased in volume as I tickled her throat and offered my congratulations. Something in her expression of tired contentment told me that she had no more kittens inside her, and so I bade her good night and stepped out into the cattery.

Although she had kept me up most of the night, Pe certainly had not wasted any time once she started to deliver; many females I had attended took twice as long or even longer. It was just past four, and with the eastern horizon glowing bright orange, the cattery was no longer dark. Many of the cats were already up and about and those who were not soon roused themselves upon hearing me scatter *niboshi* over the floor. I replenished the tins of dry cat food, swept around the sawdust tubs, replacing the sawdust in a few of them, and then headed gratefully for an early breakfast and bed.

In view of my intention, on establishing the cattery, to let the cats do exactly what they pleased within the bounds of the building, my practice of isolating females in separate quarters for childbirth and the first few weeks of kitten-rearing may puzzle readers. The fact is that to leave expectant mothers with the rest of the cats would have been to condemn most, if not all, of the newborn kittens to death and their mothers to untold misery. I knew this from the outset as a result of two tragedies that had occurred in the main house where the females concerned had not been isolated and had lost most of their kittens by the time they were discovered in the process of giving birth.

While I had nothing to do with the care of the cats at that time, I must take full responsibility for a similar tragedy that occurred in the first spring in the cattery. One morning I found the half-eaten remains of three kittens and one understandably distraught female. Either I had been careless in my observations of when she had mated, or she had given birth prematurely, but the result was that she had lost her whole litter. On this occasion only an informed guess as to the possible cause of the kittens' deaths was possible, but the culprits in the main house had apparently been adolescents of both sexes, two of whom had been discovered with still-alive but badly mauled kittens in their mouths. Rat-size and wriggling, smelling of amniotic fluids and blood, and uttering high-pitched cries, the kittens might have been a powerful prey stimulus to these young cats who had never seen kittens before and whose hunting drives, having no natural outlet in their confined quarters, would have been very easily triggered by any vaguely appropriate target.

I can, however, think of a host of other

reasons why newborn kittens stand a slim chance of survival under such conditions. For one thing, a female immediately prior to and after giving birth appears to smell very attractive to tomcats, who pester her constantly as a result. While she might be able to thwart the advances of one or two toms, keeping a whole horde at bay would severely impede her ability to give birth and take care of her kittens. Also, females, especially those with kitten-rearing experience, find their maternal instincts stimulated by the newborn kittens of other females, and they will sometimes attempt to steal kittens from the real mother. Even when they do not go this far, they often try to join in the nursing of the kittens, and if they are as numerous as they were in the Kingdom, will inadvertently suffocate all but the hardiest. Finally, one cannot ignore the mental anxiety of a mother cat trying to perform what is traditionally the most personal and private of tasks under such public conditions. Who can blame her if she simply gives up or even cannibalizes her own kittens as they are born?

If it served any purpose at all, the tragedy in the cattery confirmed the impossibility of rearing newborn kittens under such conditions and the necessity for isolating expectant females—unless, that is, one was aiming for zero population growth, and was completely heartless in the process.

Any stray mother cat who had to fend for herself while nursing kittens would no doubt have envied Pe. With warm and completely safe quarters, and all the food she could want, she was able to lead a relatively leisurely and anxiety-free life and devote all her time to her kittens. Her room was only a bare ten square yards, but she showed little inclination to leave it when I invited her to take a little exercise in the cattery. During the first week or so, she left her nest only to go to the lavatory, to eat and, when each afternoon the sun slanted into her room, to indulge in half an hour of sunbathing. Upon returning to her nest she would give all her kittens a vigorous licking, purring as she did so, and paying special attention to their hindquarters, lapping up the urine and feces that they expelled as a result of this stimulation.

At first sight Pe's kittens appeared to lead an even simpler life, an endless cycle of suckling and sleeping, suckling and sleeping. Curled up against each other in a tight little bundle, or lined up sucking at Pe's teats, they presented a picture of perfect peace and contentment. Closer investigation reveals, however, that the peace that reigns at the teat is by no means given, but is a consequence of a power struggle conducted among the kittens almost from the first time they suckle; a struggle which, put very simply, determines which kitten may suckle at which teat. By regularly noting which kitten is at which teat, as I did for Pe's kittens three or four times a day for about a couple of weeks after birth, one discovers that each kitten shows a definite preference for a particular teat, and that in most cases these preferences are decided in the first day or two of life and usually

An endless cycle of suckling and sleeping. If the kittens scuffled, Pe would roll further on her side to give them easier access.

remain unchanged for some time thereafter.

The existence of such preferences can be dramatically demonstrated by swapping the kittens around when they are suckling. Such mischief will immediately result in a brief but vigorous squabble among the kittens as they crawl over and under each other, pushing with hind legs and flailing at each other with forelegs, each one frantically following its nose to lay claim to the teat which carries its own particular smell. I knew about such "teat territoriality" long before I ever got involved with cats, but to read about such a phenomenon in a dry reference book and to witness it taking place before one's eyes are two very different things. The sight of several tiny newborn kittens, assisted only by a keen sense of smell and still rudimentary powers of movement, already so actively pitting the strength of their bodies and wills against each other in the struggle to survive, to establish and defend the first territories of their lives, never failed to impress me.

It is inevitably those kittens strongest in body and will who lay claim to the rearmost teats, the most productive ones in terms of milk flow. The runts of the litter, if they could speak, would probably object to my use of the word "preference" in their case, since they are forced by their stronger siblings to make do with the not-so-productive anterior teats. They can and will suckle at rearmost teats when they are vacant, but are liable to be brushed off easily by the

Dreaming of its mother's teats?

Pe gets an eyeful of claws. Like all mother cats, she was very long-suffering while her kittens were still small.

"owner" when he or she awakes to suckle again. Any but the weakest kittens will, however, prevail at their own suckling sites over stronger siblings, displaying in doing so rank in relation to location (relative social hierarchy) that is the feature of adult territorial behavior. In the case of a lean mother cat nursing six or seven kittens while having to fend for herself, differences in the strength of the kittens at birth would be critical to survival, kittens monopolizing productive teats winning out and growing fat at the expense of weaker siblings. There was no such worry with Pe, however; while she could never have been called fat, she ate heartily and had more than enough milk for all her offspring, whose bellies as a result always looked healthily bloated.

The teat at which I found each kitten suckling was invariably the same, no matter which side of her body Pe was lying on. She tended to switch sides after lying in the same position for some time. Such behavior ensured that the kittens suckling, often with difficulty, at their respective sites on the bottom row of teats would, after Pe had switched her position, have an easier job now that those teats comprised the top row. Whether Pe did this simply for her comfort, just as all animals do naturally in their sleep, or whether she was consciously thinking about the needs of her kittens, I don't know, but she responded to any squabbles among her suckling kittens by rolling further on her side, a move which seemed very definitely designed to help those kittens on the underside to suckle more comfortably.

For me at least, there are few things cuter than the sight of a litter of newborn kittens suckling steadily to a state of repletion, and one by one slipping off a teat and into deep sleep, little arms and legs twitching occasionally (almost as if each twitch represented another connection made, another tiny step on the road to maturity). Sometimes a kitten's mouth will remain open for a short while after releasing a teat, its tongue slightly protruding and still curled in the shape with which it had gripped the teat. If one watches carefully, one may just be lucky enough to see that little pink tongue perform perfect suckling movements *in vacuo*, almost as if the kitten is dreaming of the succour and warmth of its mother's teats, perhaps the only dream it is capable of at this early stage in its life—although heaven knows what form these teats assume in the dreams of a tiny kitten whose eyes are still shut tight on the world.

Pe's kittens were two weeks old when I first observed them rolling around to make grabs at each other's tails, legs and faces. In other words, a third dimension—play—was beginning to intrude upon the cycle of suckling and sleeping which had governed their lives up until then. Their eyes had been open for a few days and, even in the dark confines of their nest, they were becoming increasingly aware that their world consisted of more than their mother's teats and that siblings were more than warm pillows and occasional obstacles to suckling. The kittens played not only among themselves but also with their mother, Pe, who never objected to being made a punching bag by them, suffering stoically even when their needle-sharp claws caught on her nose and eyes. Only very occasionally, when she was determined to give a playing kitten a good lick-down, would she warn it to remain still by gripping the nape of its neck briefly but firmly.

In addition to the delivery room that Pe was occupying, there were two other rooms in the building known respectively as the maternity ward and the nursery. The latter I created merely by sectioning off a large corner of the near shed with another piece of Mutsu-san's golf netting, and was occupied during "kitten season" by mothers with kittens four to seven weeks old, prior to being released into the general cattery area. Here, with no more than netting separating them from the rest of the building, kittens were able to acquire an impression of the world into which they would be released, and to get to know the other cats a little, while their mothers could continue to care for them in relative peace. The maternity ward was sandwiched between the delivery room and the nursery, in the space between the two sheds, which I had roofed over with transparent plastic sheets. Mother cats were moved in here with their litters from a couple of days to a couple of weeks after birth, depending upon the "waiting list" for the delivery room. Outside the kitten season, both the nursery and the maternity ward were open to all the inhabitants.

Since in 1980 Pe was the last female to give birth, I was able to keep her in the delivery room as long as I wished, but because the maternity ward was brighter and more spacious I made it available to her two weeks after birth, as soon as I was able to transfer the previous occupants to the nursery. After preparing a new nest, I opened the hatches that separated the two rooms, settled down with my camera and called Pe. I knew that I could depend upon her to do the rest, for, as many people who have cared for a cat with kittens will know, most mother cats are born house-movers. Even if an expectant female consents to give birth in the nest one has so carefully prepared for her, within a week or two she is likely to transfer all her kittens to a bedroom wardrobe or the recesses of a broom closet, for no reason that an unenlightened owner can think of. The owner should not be offended, for the cat is with this behavior only proving what a diligent and caring mother she is.

House-moving is instinctive behavior which makes very good sense under the natural conditions in which it evolved. A mother would be wise to move her kittens every so often to make them more difficult for predators such as foxes or ferrets to find, even if she has not come across the scents of such predators near her nest. Nest hygiene is also an important consideration; I found that regularly changing the straw bedding of nests lessened the urge of many mother cats to move house. From about the age of two weeks, kittens are already able to urinate and defecate of their own accord, and will increasingly take themselves off to less well used corners of their

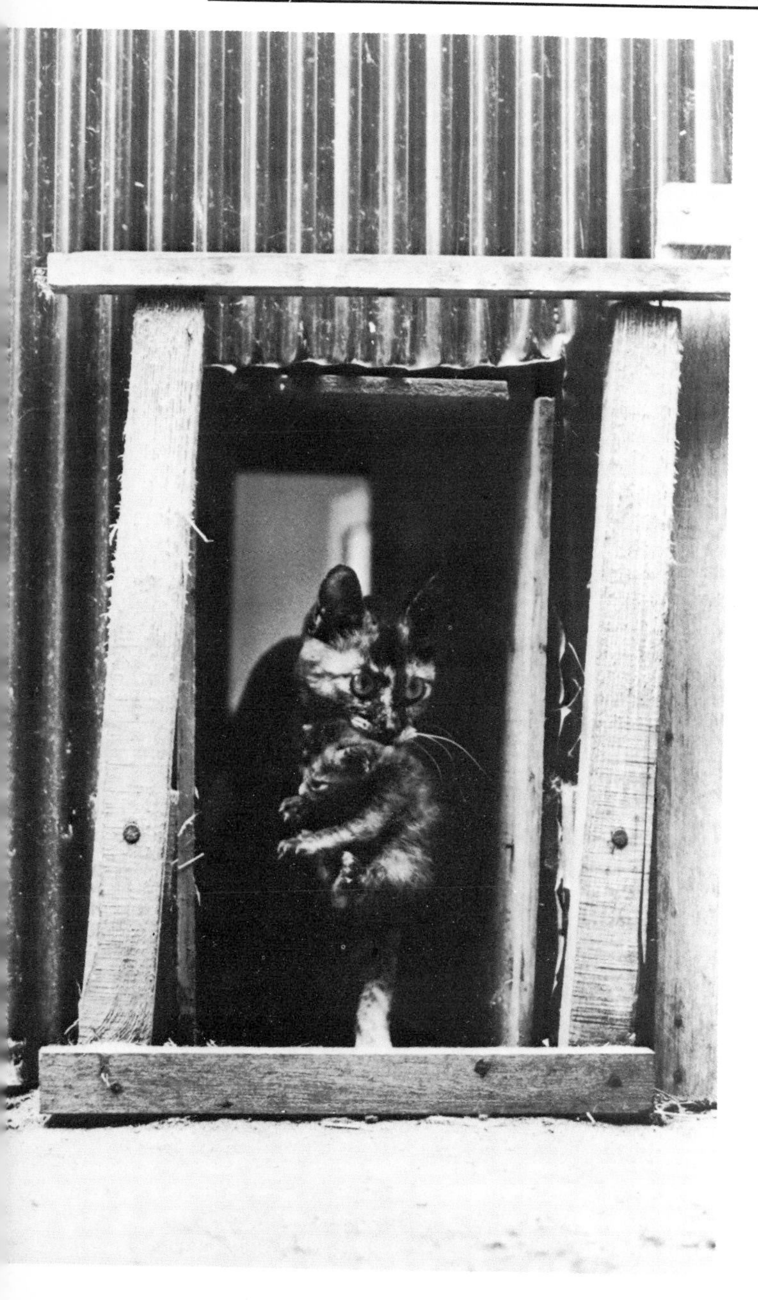

Pe carries Chestnut through to the new nest.

nest to do so. While their mother can eat or remove feces, she cannot deal with urine and, apart from the unpleasantness of wet bedding, the smell of stale urine would only add to the danger of being discovered by predators.

Pe had completed her move in a matter of minutes. Since she was already familiar with the maternity ward, she gave it only a brief inspection, paying most of her attention to the new nest that I had prepared for her. This apparently meeting with her approval, she strode purposefully back into the delivery room and returned a few seconds later carrying one of her kittens in her mouth. Depositing this kitten in the nest, she quickly returned for the next and the next, until all five were settled in the new nest. Cats cannot count, and so Pe was behaving with perfect sense when she returned twice more to her old nest to check that no kittens remained there. When she seemed satisfied that the job was done, I shut the hatch.

Transferring her kittens to an empty room, Pe had things easy. More often, females had to be shifted to quarters already occupied by others with litters of kittens. Even if the new female was accepted by the occupants, leaving house-moving to the mothers gave rise to such problems that I almost always did it for them. As a classic example, I can recall the time I wished to unite two families, the mothers of which were sisters who were on very good terms with each other and had given birth during the same period. When I opened the hatch that separated them, they each came through to the other's room, greeting each other amicably on the way, clearly recognizing each other, and then went on to investigate each other's nest of kittens. This was when the fun started. What was an "old" nest to one of the sisters was of course "new" to the other, and the two of them proceeded, at much the same busy pace, to transfer their respective kittens (coincidentally, each had four) from "old" nest to "new"; with the result that, while the two litters got thoroughly mixed up, roughly the same number of kittens remained in each nest. As the person responsible for this awkward situation, it was no doubt cruel of me to laugh, but I couldn't help myself. Humor is born out of paradox, and here was paradox with a capital P—so much industry getting the mother cats absolutely nowhere! Occasionally, one mother with kitten in mouth would stop politely before the hatch to let her sister, also carrying a kitten, pass in the opposite direction, and not a hint of puzzlement could be detected on the face of either. Once more I was being treated to the spectacle of instinct triumphing gloriously over reason and intellect, a spectacle which was as impressive as it was comical.

I thought that time would sort the problem out, especially when one of the mothers settled in what was to her a "new" nest, but no such luck; she leaped back into action the moment her sister appeared to carry off another kitten. Perhaps if I had left the two of them to their own devices, a solution might have been reached by nightfall, but after watching them for an hour, I had had enough. While their instincts and energies appeared to be inexhaustible, my well of laughter had long since dried up. Compassion for the kittens more than for their seemingly unperturbed mothers got the better of me, and I lined a fresh cardboard box with straw, placed both families inside, dropped the hatch between the rooms and left the sisters to come to terms with this simpler situation, something which, to my relief, they did very quickly once they realized that they had no alternative.

Pe's kittens started to venture from their nest at around three weeks and, within a week, were thoroughly familiar with every square foot of their quarters. The sight of them using a proper litter tray with increasing frequency was a very welcome one, since I was soon relieved of the necessity of changing their straw bedding every couple of days. With her kittens becoming more active and disobedient by the day, Pe had little choice but to let them come and go as they pleased, but when they were not in the nest she would watch over them, always ready to leap between them and the wall which separated them from the dogs outside when she heard the latter stop to sniff immediately on the other side as they passed by. Normally Pe and the other cats paid little notice of the dogs, who could not possibly get into the building, but now Pe was leaving nothing to chance. When she heard the dogs approach, she would shoo or call her kittens to the other end of the room, in the process teaching them to beware of anything that smelled or sounded of "dog."

To venture or not to venture . . .

At the age of one month, the kittens were already helping themselves to Pe's fish and rice, and Pe seemed only too happy to let them. Mother's milk became more and more a luxury rather than a necessity, a luxury that Pe was increasingly unwilling to proffer to the kittens. Watching the way they went at her teats now, with a vigor that threatened to tear them out by their roots, one could hardly blame Pe for the intolerance that she sometimes showed.

The focus of my interest in cats was their social behavior rather than their hunting behavior, and so I did not go out of my way to provide them with a regular supply of rats. But for their entertainment, and because I was keen to give kittens in particular a taste of the kind of thing they were designed by nature to do, at an age when they were most sensitive to the stimulus of prey (many of our older cats who had had no exposure to prey in their childhood showed absolutely no interest in rats), I regularly set rat traps in the stables and outhouses. The traps were of a cage variety which capture the victims live, and I gave adult rats to the adult cats, and young ones to kittens. Unfortunately, most of the victims were real monsters, often more than a match for all but the bravest of our rat catchers, let alone kittens, but one day approaching mid-July, when Pe's kittens were just over five weeks old, Chiyoko burst into Pe's room with the news that she had found a youngster in one of the traps—happy news for Pe's kittens, if not for the poor rat!

Mother cats play a very important role in the hunting education of their kittens, and so I never excluded them when I had a rat for their kittens. After allowing Pe and the kittens to sniff at the rat while it was still in the trap, we released it. It was Pe's first rat for a long time, and she showed her excitement by leaping on it and killing it in an instant, giving the kittens no chance whatsoever even to see it run. Pe squatted motionlessly facing her kittens, the rat firmly in her mouth and a fierce expression on her face. A kitten approached to investigate but was sent scuttling backward by the low growl that Pe suddenly uttered. It was the first time that the kittens had heard such a sound and they reacted with surprise, but they seemed to grasp its meaning quickly, for they all retreated up to the nest as if they knew Pe wanted to be left alone.

She stuck in the same spot for two minutes, the rat still in her mouth, and still very excited, growling occasionally. However, her attitude suddenly changed; although she was still tense, the ferocity left her eyes. She dropped the rat and called to her kittens, a call which, while beckoning, was clearly different from the call which meant "milk," or any other call that the kittens were familiar with. I had witnessed this situation a few times before, and recognized the call immediately as the one meaning "prey." After facing this situation once or twice, kittens, too, soon learn the meaning of this peculiar call and come running instantly, but Pe's kittens were

In response to Pe's call, Raspberry approaches the rat, still unsure of herself.

Frozen with concentration, Pe watches as her kitten plays with the rat.

hearing it for the first time and, in view of Pe's somewhat frightening attitude up to that point, seemed to doubt its beckoning tone.

On this occasion it so happened that they had reason to be hesitant, for Pe seemed in two minds as to what she wanted to do. She remained still as one of her kittens began to approach, but then suddenly grabbed the rat again and growled, stopping the kitten in its tracks. No sooner had the kitten started to turn away than Pe dropped the rat and called again. Her expression was complicated. It seemed to me that two instincts were battling each other within her; one, the instinct of a good mother to educate her kittens, the other, the "finders-keepers" instinct of a solitary hunter. Having killed for the first time in a long while, the latter instinct appeared to be so strong in Pe that it threatened to override the former, but in the end it was the mother in her that won the day. The next time the kitten approached, she remained silent and still and let it sniff the rat beneath her nose.

After inspecting it thoroughly from head to tail, the kitten sat back, pricked its ears forward and tentatively stretched out a paw. The kitten's first cuffs at the corpse were very gentle, but soon they had escalated into full-blown swipes with unsheathed claws, shifting the corpse this way and that, and occasionally skyward, whereupon the kitten would grab it with both paws, pulling it to its mouth to deliver a nip or two before letting it drop again to the ground. Pe sat up stock still behind the kitten, her eyes glued to the rat in fierce concentration.

When, after five minutes of this, her kitten showed signs of losing interest, Pe stretched out a paw to flick the corpse around, bringing it to life, so to speak, to stimulate the kitten to further play. Two other kittens, who had been watching their sibling's antics as if entranced, approached and sniffed the corpse, and then one of them leaped into action, tossing it, pouncing on it, occasionally biting it quite powerfully and growling at any of its siblings who approached. Kittens are never halfhearted in their play, but it was clear that the rat was not just another toy. It was triggering far more intense emotions, stirring for the first time the very core of the "hunter" in the playing kitten. For some reason, two of the kittens appeared to be not yet susceptible to such a stimulus and showed only passing interest, but the other three took turns at pummeling it for more than an hour. Pe left them to it, but continued to watch from a distance with a look of approval on her face.

Among the rats we caught in the following two weeks were another two young ones. Pe again killed the first of these instantly, but this time called her kittens immediately and left the rat to them. On this occasion, all five kittens participated in "training," two of them progressing to nibbling at the hind legs of the corpse. Pe herself, like many of our cats, had never shown a liking for the taste of rats, although she killed them readily enough. Hunting and eating are clearly two separate and independent drives, and necessarily so. Marvelous hunters though they are, cats will meet with failure far more often than success during hunting expeditions. It therefore pays them to be ready to kill whenever the opportunity presents itself, and not just when they are hungry.

The kittens got their third rat a few days later, by which time they were occupying the nursery. At first sight I thought that Pe had killed this one as quickly as she had the first two; it certainly looked very dead as she held it in her mouth and called her kittens. But when she released it, I noticed that it was very much alive and unscathed, even if it appeared to be in a state of semishock. The ability of cats to catch prey alive and the practice of mother cats to present their kittens with live prey at a certain stage in their hunting education are well known, but here again, actually watching a cat chase and capture its target with all the speed and precision it would use when bent on killing and yet hardly scratching the prey,

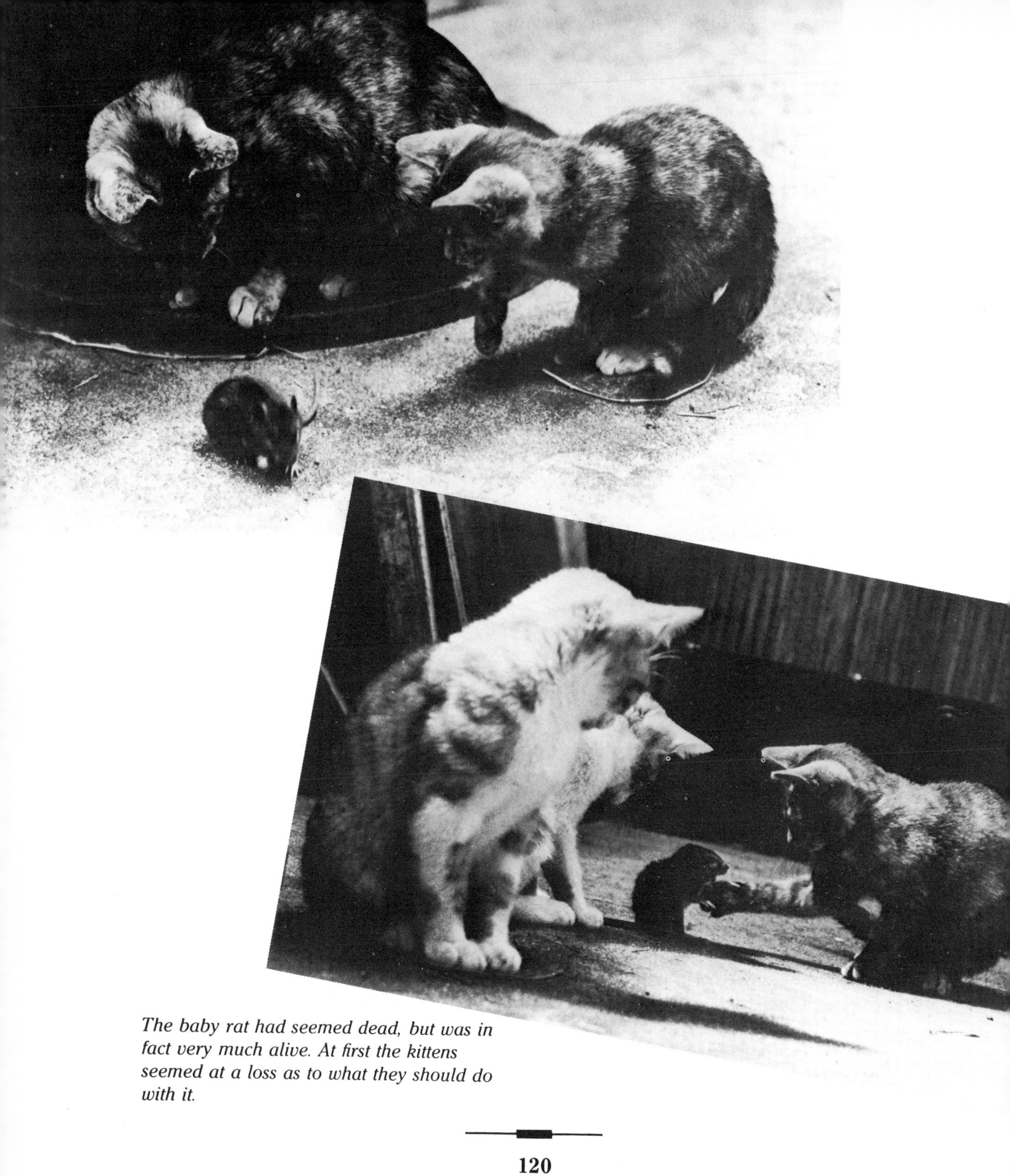

The baby rat had seemed dead, but was in fact very much alive. At first the kittens seemed at a loss as to what they should do with it.

brings home to the human observer, in a way that no written account could, the marvelous nerve coordination possessed by these animals. The sight simply awed me every time I witnessed it.

Pe clearly considered that the kittens were ready to tackle live prey, but not any live prey. Later that day I released a much larger rat, caught at the same time, into the nursery. Pe was an efficient rat catcher but not among the bravest, usually showing considerable trepidation at going after "monsters." That day, however, concern for the safety of her kittens clearly girded her with courage, for she dispatched the monster I had released without the slightest hesitation.

Pe's kittens seemed to be somewhat at a loss as to what to do with the live young rat, just staring at it intently as they surrounded it. The immobility of the youngster did not help them to spring into action: a moving target is much more stimulating to cats than a stationary one. Apparently realizing this, Pe extended a claw and tapped the rat, making it move a few steps. Soon her kittens were following her example, although with restraint, unsure of themselves. The rat was still in a state of shock, and not yet given to run. When tapped, it occasionally reared on its hind legs to face its assailant before moving again. The kittens followed it around for about five minutes at a leisurely pace, eventually making it jump into their tin of cat food.

To my amazement, it picked up a pellet in its forefeet and started to nibble. Since it might have been in the trap most of the previous night, there was a good chance that it was hungry, but its sudden display of appetite was almost certainly also displacement activity, brought on by bewilderment at its situation. Anyway, its nibbling seemed to have the effect of injecting it with a little more life, for when one of the kittens next cuffed it, it leaped out of the tin and this time ran with some vigor. It was this sudden burst of speed that sealed its fate, for the kittens began to show real enthusiasm for the chase. And then, when one of the kittens disappeared under a low shelf in pursuit of the rat, it was all over. Crouching to get a better view, I found the kitten with the rat already in its mouth, growling at its siblings to keep away. This time, there was no mistake that the victim was very dead. In the kitten's eyes was the same fierce look that I had observed on Pe's face when she killed the first rat I gave to the kittens. The kitten had killed for its first time, and clearly knew it.

Pe and her kittens had been in the nursery for a week. Since moving them, I had been letting Pe out twice a day, and she had gone willingly, although she always registered her desire to get back to her kittens after wandering around for an hour or so, by hanging around the nursery door and calling for someone to let her in. Her kittens were now almost seven weeks old, and big enough to take care of themselves in the cattery, and so, the day after they, or at least one of them, had been initiated into the art of killing, I opened the door to the nursery and let them enter the adult world that lay on the other side.

Kagetora, "Shadow Tiger," in the days of his young adulthood, when his brash self-confidence more than emulated that of his samurai namesake.

Kagetora safe and warm, but lonely, in his box below the rafters.

The cattery's first, and so far only, permanent pariah. From his impregnable pillar-top position, Kagetora resists pressure from Kenbo. Note the scar on his hind leg—the insignia of a pariah.

Ownership of prey was a matter where possessiveness overrode considerations of rank.

Chiyoko, her feet submerged in a sea of cats, doles out their fish.

Ganko scent-marking a branch on one of the trees in the cattery.

Mealtime in the cattery, a democratic affair.

A cat established in a favorite spot would often stand his ground against a senior. Here Sha looks straight through Mo as the latter threatens him.

Three generations: Pe, her daughters Chestnut and Plum, and their combined litters.

Welcome to the world.

10

ASPECTS OF MOTHERHOOD

Nineteen eighty was baby boom year in the cattery, with a constant stream of females giving birth from the beginning of March through to June. Even with two postnatal rooms, I had no choice during most of this period but to house females with litters of kittens of approximately the same age in the same quarters, even if they objected. It was thus circumstances rather than any intentional experimentation on my part which led to the discovery that in most cases females with kittens not only put up with sharing living quarters but almost seem to prefer to do so. I was convinced that, no matter how peace-loving and amicable the females were toward each other at other times, they would display a very different attitude as mothers, since mother cats are famous for the manner in which they will attack any intruder that they consider poses a threat to their kittens. Even if a strange female posed no such threat, I did not expect her to be made to feel welcome. My apprehensions were justified in a few cases when a certain female proved so aggressive toward another introduced into her quarters that I had to subdivide the rooms further or remove the troublemaker and her kittens to our rooms in the main house. In view of such problems, you can imagine my delight, not only from a scientific but also from a logistical point of view, when the large majority of mother cats settled down peacefully to rear their kittens communally.

Six mothers, eighteen kittens—a communal nursery in which the kittens seemed to thrive . . . "The more the merrier" seems to be the motto of most mother cats.

Even if I could not provide every female with a separate room, I took care to give each a nest to herself. These efforts, however, proved worthless; no sooner had I transferred a cat and her kittens to a separate nest in the maternity room than the other females occupying the room would come to investigate, and immediately carry off the kittens to their own nest. Far from objecting to such behavior, the newly introduced female would, in her own investigations of the room, come across the other females' kittens and settle down with them, almost as if she considered them more attractive than her own. In this way, despite my provision of about six nests in each room, two or more families would very quickly fuse into one, with the mother cats caring for their own and other kittens indiscriminately. This tendency to pool litters forced me hastily to construct "apartment complexes," large boxes subdivided into separate nests by boards which could be removed to create larger nests as the situation required and reduced the danger of weaker kittens being suffocated. Even in single-litter situations, a mother cat will occasionally inadvertently suffocate a weak kitten which, with its breathing impeded, fails to struggle free or draw its mother's attention by crying. This danger is, however, much greater in multilitter situations where, even given ample space, the mothers and kittens all tend to clump together in one big pile.

Almost all the kittens in these situations, however, not only survived but lived a life of luxury. At one time, I had six mother cats nursing eighteen kittens in one large nest. Since the mothers nursed the kittens indiscriminately, "teat territoriality" broke down completely, each kitten helping itself to the nearest available teat when hungry. Even if we count only six teats per mother (since the front two teats are not very productive), there were always teats available whichever way a kitten turned, and milk was permanently "on tap." Moreover, with six tongues working around the clock on them, these kittens were probably the best-groomed members of the cattery. Friendly scenes, such as one female nursing three kittens while licking the rear ends of two more suckling on another female who was in turn exercising her tongue on another kitten or two, were commonplace. And with so many cats gathered together, the nest was always as warm as toast. It was hardly surprising, then, that the kittens appeared to flourish.

Some readers may wonder why mother cats will, in almost all cases, nurse any kittens, irrespective of whether they are their own or not. Insofar as the purpose of reproduction is to ensure the proliferation of one's genes by populating the earth with one's descendants, spending time and energy caring for unrelated kittens would appear to be a distinctly counterproductive activity. As with many other puzzling aspects of the behavior of domestic animals, this one also can be largely answered by pondering upon the kind of life that the animal led prior to domestication.

In all likelihood, females of the domestic cat's wild ancestors gave birth and reared their kittens alone. There might have been no other females with kittens for miles around. Under such circumstances, a mother wild cat could be absolutely certain that any kittens in the vicinity of her nest were her own; there was simply no need for her to be able to distinguish between her kittens and others. In the absence of evolutionary pressures to change in this respect, today's domestic cat shows the same "blindness" with regard to kittens, and it is this blindness, coupled with strong maternal instincts, which will cause a mother cat to attempt to nurse umpteen kittens given the opportunity, and even baby animals of other species. I once saw a photograph in a Japanese newspaper of a cat nursing two tanuki (racoon dog) pups along with her own three kittens. Despite their superior size, the tanuki pups were clearly suckling on anterior teats, with the kittens firmly glued to the posterior ones—if this was not just coincidence, it says a lot for the determination of the kittens!

Far more intriguing from the behavioral point of view than the females' acceptance of any kittens was, however, their acceptance of each other. As mentioned earlier, certain females objected from the start to sharing quarters with each other, particularly when their kittens were still very small. Moreover, I cannot say that things always went smoothly afterward between females nursing their kittens together. For example, in the case of the six mother cats, I did not once see any squabbles occur inside the nest; but outside it, away from the kittens, there were occasional cross words, hisses and paw blows exchanged between the females. Communal kitten-rearing was clearly not a completely tension-free situation.

I noticed, however, that squabbles rarely occurred among females who were closely related and knew themselves to be so. In 1979 and 1980 the sisters Uko and Aya gave birth to kittens within a week of each other, enabling me to unite them and their kittens soon after the birth of the second litter. On both occasions they settled down together immediately and there was never a hint of tension between them. They seemed, in fact, positively glad of each other's presence. The same atmosphere existed between two more sisters, Chi and Toco, when they, too, reared their kittens together.

One more case involved three cats. In 1981, Pe and both her daughters of 1980, Plum and Chestnut, gave birth within two weeks of each other. Chestnut was the first to give birth and proved to be an able mother, but she was overjoyed when I united her and her kittens with Pe ten days later, five days after Pe had given birth. Plum was the last to give birth and, right from the start, showed no signs of nursing her kittens. She did not try to cannibalize them, but seemed highly excited and just would not settle down. Despite the small size of her three kittens, I took a gamble and transferred the family to the nest occupied by Pe and Chestnut. Plum's attitude changed immediately. Nestling in with her mother, sister and their kittens, she began to purr loudly and devote herself to all the kittens with the same diligence shown by Pe and Chestnut. We removed her smallest kitten for hand-rearing, fearing that it was losing out in the competition, but her other two kittens proved strong enough to survive, which they would not have done if we had left Plum to care for them alone.

How are we to interpret these findings? Were the females who did cooperate in kitten-rearing merely making the best of an awkward situation, maintaining peace among themselves for the sake of their kittens? The occasional squabbles among the mothers nursing their kittens together certainly seemed to suggest this at times. However, even unrelated mothers got on well with each other most of the time. And the atmosphere that existed between Pe and her daughters, between Uko and Aya, and Chi and Toco, when these females were nursing their respective pooled litters, was so warm, relaxed and cooperative that it was hard not to conclude that pooling of litters and assisting each other in kitten-rearing was completely natural behavior, at least where closely related females were concerned.

Particularly in a species in which the males play little or no direct role in the care of the young, the advantages of cooperative kitten-rearing are quickly apparent. Two or more sets of eyes, ears and claws provide the kittens with better protection than one. Moreover, if two females shared maternal responsibilities, one mother could remain with the kittens while the other went off to hunt; there would be no need for the kittens to be left alone as in a "single parent" situation. Even if no blood ties existed between the females involved, such advantages might well outweigh the disadvantages of caring for unrelated kittens in addition to one's own. However, it is likely that in a natural situation females rearing their kittens cooperatively would most probably be related as sisters or as mother and daughter. In such cases each female would be related to kittens other than her own as either aunt, grandmother or half sister.

In other words, although a female's own kittens would have most in common with her genetically, she would share a lesser, but nevertheless significant fraction of her genes in common with the kittens of her mother, sisters or daughters. Bearing in mind that the aim of reproduction is to ensure the proliferation of one's own genes in a population, a female would naturally have a vested interest in the survival not only of her own kittens but also of other related kittens. The more closely related two females were, the more they would have to gain by the practice of cooperative kitten-rearing, and even a female without kittens of her own would, by assisting in the upbringing of related kittens, be working in her own reproductive interests to the extent that these kittens possessed genes in common with her.

The behavior of the females in the cattery seemed to fit in very neatly with this argument. Uko and Aya were full sisters who had come to the Kingdom at the age of six months, had shared the same quarters in the main house and had continued to display a warm attachment to each other after transfer to the cattery where they could often be seen sleeping together or eating at the same tin of cat food. Toco and Chi were two of the litter of five kittens that Chiyoko and I hand-reared from the day of their birth. We had put them through all sorts of adventures as kittens, and they continued to be the greatest of friends, even as adults in the cattery. As for Pe, the instance described in Chapter 5, when she answered the distress calls of seven-month-old Cherry, a full brother of Chestnut and Plum, showed that she still clearly recognized her offspring and was willing to come to their aid long after they had become independent of her.

The females in these cases were, then, very much aware of their relationship to each other, and the ease and lack of friction with which they reared their kittens together suggested that female cats do indeed possess a natural inclination to coexist peacefully and actively cooperate with close relatives. The fact that almost all of the cattery house females got on well with each other, and that unrelated females also cooperated in rearing their kittens, in no way undermines this argument, for it is perception of relatedness rather than actual relatedness which is important. In a normal kitten-rearing situation, such perception develops naturally between a mother and her offspring, and among the offspring themselves as they grow up together, but it can also develop among unrelated individuals as a result of long association, particularly if this association begins when the individuals concerned are still young.

Almost all the unrelated females who cooperated in kitten-rearing in the cattery were fairly young and had spent most of their lives together. Some had played together as kittens. Sleeping together and grooming each other daily, they probably felt as much related to each other as any genuine siblings do. Their warm relationships with each other and their cooperation in kitten-rearing only served to convince me further that, at least for female cats, a social existence might be as natural as a solitary one, if not more so. While the cattery population was an artificial aggregation, the like of which would never be encountered naturally, my findings there nevertheless suggested strongly that in a natural setting, with an adequate food supply, cats would form small but stable groups, the core of which would consist of two or more closely related females. While the members of the group would spend a good portion of each twenty-four hours hunting alone (the hunting of small prey is necessarily a solitary activity; one mouse is a meal for only one cat, and two cats going for the same target would only get in each other's way), they would also spend time together in a shared home base, finding comfort and support in each other's company, cooperating in kitten-rearing, defending a common territory. In other words, they would comprise a coherent social unit, not just a loose aggregation.

In 1982, when I first put these ideas down on paper in a Japanese version of this book, I had virtually no information on any other studies which might provide support for my conclusions. Professor Leyhausen, in his book on cat behavior, makes no men-

The sisters Aya and Uko with their kittens—never a cross word.

Having spent most of their lives together, grooming each other daily, most young females probably felt as closely related to each other as genuine siblings.

tion of the phenomenon of cooperative kitten-rearing. Although he does credit domestic cats with a certain degree of sociability, he tends to stress the solitary side of their natures, and appears to be of the opinion that, while cats can, when forced to, live together quite peacefully, if given the choice they would opt for a solitary lifestyle.

It was shortly after my manuscript for the Japanese book had gone to the printers that I received, by way of my mother in England, a book called *The Curious Cat*, which described a detailed, scientific study of the behavior of a group of four cats, a tom and three females, over the period of a year on a Devonshire farm to which they had been brought in early 1978.* Any reader familiar with this book or with the BBC television program about the study will be able, I think, to appreciate the joy and relief that I felt when I read it, since the behavior

*Michael Allaby and Peter Crawford, *The Curious Cat* (London: Michael Joseph Ltd., 1982). This study was sponsored by the BBC with the aim of producing a program on the behavior, particularly the social behavior, of ordinary cats who had to fend for themselves in a natural environment. One of the coauthors of the book, Peter Crawford, was the producer of this program. The help of zoologists at Oxford University was enlisted to carry out the scientific observation and research on the cats. Supervising this side of the project was Dr. David Macdonald, a contemporary of mine in the same undergraduate class at Oxford, who went on to obtain his doctorate (on the behavior of foxes) and become a member of staff in the zoology department. Even more of a coincidence, Peter Apps, the young zoology graduate whom Dr. Macdonald recruited to carry out all the observation and data collection on the cats, attended not only the same Oxford college, St. Peter's, but also the same grammar school as myself, being seven years my junior. It's a small world!

of the females in the study conformed in so many ways with the picture suggested by my cattery observations of the way of life that might be led by a small group of females fending for themselves in a natural setting.

It would appear to be high time, at least where female cats are concerned, to revise the traditional and popular view of the cat as being an exceptionally solitary and antisocial animal. Whether cats will opt for a social existence or not will depend upon all sorts of factors, such as food distribution and individual temperament. What does seem clear from the study mentioned above and my observations in the cattery is that cats do in general possess a potential for cooperating with each other, and if the circumstances are right, may lead a warm and fruitful social existence as a matter of choice.

The first steps of kittens just released from the nursery into the cattery were understandably marked with diffidence. They would find themselves not only on the brink of a world many times the size of the one they had just left but also surrounded by a host of curious adult cats with whom they had had up until then only nose contact through the nursery netting. Although the attention that the adults paid them was rarely unfriendly, most kittens would crouch low, fluff up their fur, and occasionally hiss at any adult who was inspecting them too closely for their liking. Not the kind of characters to be discouraged by such threats, some toms would conclude their inspection of a little newcomer by spraying it. One or two, noted for their sexual appetites, namely Nibu and Panda, would even occasionally attempt to mount these kittens. Within half an hour, however, their interest would wane and such perverse antics would give way to the much more wholesome sight of the kittens, less fearful and more excited with every passing minute, bursting spontaneously into play as they bumped into each other on their initial explorations.

Within a couple of days they would already be familiar with the furthest corners of the cattery, and could be seen pounding hell for leather one after another along the aerial walkways, scrambling up and slithering awkwardly down the trees in the garden, ambushing each other and unsuspecting adults from newly discovered hideouts, wrestling in the grass and over old sofas, and in general treating the place as if it was a playground designed purely for their entertainment. Only when their energies were almost totally spent would they finally heed the calls of their mothers, whose urge to nurse, although on the wane, would still be strong enough to prod them to round up their kittens whenever the latter did not appear of their own accord. Many kittens were just as likely, however, to ignore their mother's calls, settling instead among a heap of other cats on a sofa, looking every bit as much a part of the establishment, only a day or two after joining its ranks, as any long-term inhabitant. Wherever they came to rest, they would very soon slip, with tired little legs twitching involuntarily, into sleep as deep as their play had been furious.

In view of the ease and speed with which kittens adapted to life in the cattery, my

A wrestling match on the old sofa.

only immediate concern upon releasing them from the nursery was that they got enough to eat. Some kittens would already have taken to eating dry cat food while still in the nursery, but many still found it difficult to crunch, and anyway showed a distinct preference for fish and rice. In the nursery they had as much of this as they could eat, and in any case mother cats would normally allow their kittens to eat their fill before helping themselves. Upon release from the nursery, however, kittens would soon discover that most adults would show them no such consideration at mealtimes. Although they quickly became as adept at wolfing as any adult, it was clear that, without countermeasures, they would either have to increase their intake of dry cat food or go short.

I saw no advantage and certain risks in forcing such a choice on them, and so set up what became known as "Kitty Café." This consisted of a spacious wooden box, the entrances to which were wide enough to let only young kittens pass through, and inside which I placed a large bowl of fish and rice. It was largely as a result of this idea that I discovered exactly how narrow a space an adult cat can force its way through. Time after time our adults disproved the popular notion that cats will only enter a space wider than the distance between the extremities of the whiskers on each side of their faces. When it has the will, a cat makes determined attempts to find its way into the narrowest of spaces, normally by twisting its neck and pushing its head in sideways. If this is accomplished with relative ease, most cats will succeed in forcing the rest of their bodies through.

The box started off with entrances about three inches in width, but I was forced to narrow them, quarter inch by quarter inch, every time that I found an adult feasting inside. These adults, generally slightly built females, like Uemi and Papu, with voracious appetites and a dislike of dry cat food, predictably resented such revisions, and would even go so far as to try to widen the entrances with their claws upon discovering that they could no longer get their heads inside. As a result of their determination, the entrances were in the end narrowed down to a little under two inches in width, allowing kittens of up to three months to enter with relative ease (although leaving with a full belly required a little more

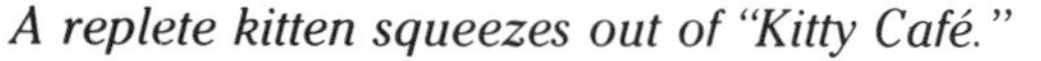

A replete kitten squeezes out of "Kitty Café."

Kitten Ping-Pong—but from the kittens' point of view it was piggy-in-the-middle. Mac, the little blue tabby, was the star performer.

"push") but increasingly difficult for kittens over that age. I located the box strategically close to the nursery entrance so that it was invariably one of the first discoveries made by kittens upon release into the cattery. Upon waking, stretching and extracting themselves from a pile of other cats, most kittens would make their way there for a bite to eat if anything remained, before embarking on another bout of play and exploration.

Because of their boisterousness, impudence and disregard for standard feline etiquette when at play, kittens occasionally incurred the wrath of their elders, but in general adults, toms as much as females, showed them a great deal of tolerance and even warmth. The high spirits of the kittens would infect young adults in particular, prodding them out of their lethargy on hot summer days to participate with the kittens in games of chase and wrestling bouts. Even staid older cats were not immune to their influence, and I occasionally witnessed Mo or Ee, as kittens whizzed past them, suddenly crouch low and then burst uncharacteristically into a run in mock pursuit of them.

Neither were we humans always content just to watch the antics of the kittens, pleasurable though this pastime was. There was an old table tennis table in the far shed, and kitten Ping-Pong soon became an established sport in the Animal Kingdom, providing us with some of the most enjoyable memories of our lives with the cats. The game in fact resembled piggy-in-the-middle far more than it did table tennis, with the kittens as piggies gathered on both sides of the net, doing all in their power to foil our attempts to get the ball across to each other. The game was thus one of man versus cat, with two players comprising the human team, and as many cats as wished to play making up the feline side. The point of the game, as far as we were concerned, was to keep rallies going as long as possible. Any rally of over four strokes constituted a point in our favor; any over nine gave us the game, this however being purely theoretical, never occurring in reality. Any rally that our opponents put a stop to before it reached five strokes was of course a point to them. The only other rule was that humans would not use their bats to clear the deck of their opponents or otherwise cause them physical injury. Even inadvertent contact between a bat and the skin of a feline opponent would cost the humans five penalty points; this ruling was necessary to prevent overzealousness on the part of the humans, something which it was easy to give in to in the heat of the moment, especially when the stroke count was on four!

It was always the kittens that hogged the centerfield. Adults would gather on perches near the table to watch the action, or sit on the sidelines, helping their team through passive obstruction. One or two occasionally provided even more indirect support by scrambling up onto our shoulders and disturbing our physical and mental equilibrium.

Any cat compares favorably with most other animals when it comes to agility and speed of reflexes, but this game demonstrated the considerable differences with regard to athletic ability that exist between individuals of the same species. Every year would see the emergence of a new star or two whose skill and enthusiasm would set them apart and make some of their teammates look like real sluggards. Of these stars, the most memorable was Mac, a beautiful little blue tabby born in the baby boom of 1980. She would leap skyward for the ball when most other kittens just reared up on their hind legs, and she was the only one who succeeded more than just occasionally in catching the ball in midflight with both forepaws. Moreover, she never let herself be distracted from the target, unlike many other kittens who would often divert their attention to games of footsy with each other under the net. Mac retired to the sidelines after one season, but she went down in cattery history as the Junior Star of all time.

The game was excellent practice for top spins, which, combining speed with height, gave the kittens most problems. Even after mastering this stroke, however, we were no match for the kittens and never once chalked up a victory, although defeat never detracted from the fun. The kittens appeared to be practicing player rotation, some taking over while others rested and, as a result, we always tired long before they did. The victors would then get the ball to themselves and, transferring to the floor, switch games from piggy-in-the-middle to soccer, chasing the ball all around the cattery until they, too, were ready to drop. There's no doubt about it, kittens are great sports lovers.

The cattery, rarely dull at any time, was nevertheless always a brighter place for the presence of kittens. I knew, however, right from the start of the project, that they were a luxury which we could allow ourselves for two years at the most, barring unforeseen circumstances. While the cattery could comfortably hold the hundred and twenty cats which inhabited it at the peak of their population, five hundred cats were a completely different matter, and this was the approximate number that I calculated would cram the building by the summer of 1982, based on the population at the end of 1979. Such crowding would not only have posed nightmarish management problems for us but would have been hell for the cats as well. For this reason, in the spring of 1980 I embarked on a female spaying program designed to keep the population within manageable and comfortable limits. I started with females who had a record of repeated miscarriage or failure to care for their kittens, going on to the remainder of the females after they had all but finished rearing their kittens of that spring and before they had a chance to get pregnant again. Their female kittens were in turn operated on early in 1981, before they, too, became pregnant.

The cooperation of our resident vet, Iwase-san, was invaluable in carrying out this task. Iwase-san had come to Hokkaido at Mutsu-san's request, primarily to manage the Kingdom's population of draft horses and Hokkaido ponies, but this never stopped him from coming willingly to our aid in the treatment of any illness or injury among the cats, and he always took pains to teach us treatment procedures whenever appropriate.

Women readers may find complaint, as some visitors to the cattery did, with my selection of the females rather than the toms as the target of our program. Believe me, it was cool, objective reasoning and not male chauvinism that led to this choice. For a start, there was little point in doubling our labors by operating on both sexes. Moreover, I wanted dramatically to reduce the reproductive potential of the population without eliminating it altogether. The latter would have been the result if I had chosen to operate on the toms, since just one sexually intact and active tomcat could have done almost as much damage among the female population as fifty toms, and we would thus have had to doctor every last one of them. Even though the females were the scapegoats of the program, we chose to have fallopian tube cuts performed on most of them rather than ovariectomies. As a result, while they could no longer become pregnant, they were able, with their ovaries still intact, to enjoy as full a love life as they had before—fuller, in fact, since their estrous cycles were no longer interrupted by pregnancies. This in turn meant, of course, that the toms also benefited enormously from the program in terms of sexual opportunity.

Even if I had sound, objective reasons for operating on the females rather than the toms, I have to admit to some favoritism in my choice of Pe, Toco and Chi as the females who, at least for one more year, would escape the scalpel. By the autumn of 1980, they were the only mature females left in the cattery with their reproductive capabilities intact. Although I regretted having to deprive the females of the joy most of them found in rearing kittens, I had no doubts whatsoever about the correctness of my actions. In a natural situation, availability of food, predation and disease ensure control of animal populations, but in a man-made situation, it is the responsibility of man to assume, in the most humane way possible, the role of nature.

Since I started my program on the females at the beginning of 1980, by the time that the first kittens born that year were being released from the nursery, there were already a number of doctored females in the cattery. Among them were Papu and Pe's sister Nan, both of whom had been operated on after miscarrying, and also Kerompa, who had a history of cannibalizing her kittens. These cats would never again raise their own kittens, but this did not stop them that spring from suddenly assuming the role of foster mothers to young kittens just released from the nursery. Since cases of animals taking on the care of baby orphans, even when not nursing their own young, have been recorded in various species of mammals, the appearance of this behavior in the cattery, heartwarming though it was, did not particularly surprise us. There is something in the juvenile features and general helplessness of baby animals which will stimulate the maternal instincts of adult females (and males), even of unrelated species. Konrad Lorenz mentions some such cases in his marvelous book, *King Solomon's Ring*. And the memory of one of our pack of dogs, a bitch named Col, gluing herself almost twenty-four hours a day to a sickly draft horse foal, licking it gently and growling at any other dogs to keep their distance, will remain forever dear to me. What really amazed us in the cattery cases, however, was that the foster mothers were, within two or three days after starting to nurse the kittens, producing genuine mother's milk, and in quantities which soon compared favorably with a real mother nursing newborn kittens!

Female cats and dogs undergoing pseudopregnancy, feeling "broody" with their teats swollen and painful, are especially liable to seek out baby animals to nurse, but conversely, the presence of babies can bring on such a condition—or at least this appeared to be what was happening in the cattery. A month had passed since Nan and Papu had been operated on, and the onset of their maternal behavior coincided exactly with the release of kittens into the cattery. (The fact that Kerompa, of all cats, showed this behavior, suggests that the cause of cannibalization of kittens by their mother may be anxiety at birth rather than lack of desire to mother kittens.) The physical stimulus of the kittens sucking the teats of the foster mothers was no doubt vital in encouraging the flow of milk; this applies to genuine mothers as well, milk flow soon

"I said no!" Poppy lets one of her kittens know that she doesn't want him around anymore.

Papu "rescues" one of her foster kittens from a tree. More than once I saw her drop a kitten by accident.

drying up if this stimulus is absent. However, it is generally assumed that a female mammal has to go through pregnancy and give birth in the first place in order to produce milk. Indeed, the hormone oxytocin, which stimulates birth contractions, also plays an important part in the stimulation of milk flow. Although wet nursing has been practiced from ancient times in almost all human societies, I have never heard of cases of women producing milk without giving birth themselves.

The foster mothers in the cattery were perfect mothers in other respects, too, attending to the kittens with far more diligence than their real mothers were by that time. The latter, with the bulk of their maternal obligations fulfilled, would usually begin to rebuff the approaches of their kittens, hissing and cuffing at them when they tried to suckle. The fact that mothers of single kittens will continue to suckle them for much longer periods suggests that mother cats find it harder and harder to cope with the increasingly aggressive sucking of a number of kittens on their increasingly unproductive teats. Even if the mother's reasons are largely selfish, such rebuttal no doubt has the beneficial effect of encouraging "hanger-on" types among the offspring to become more independent and start to fend for themselves.

In the cattery, however, such kittens merely switched their attentions to the foster mothers who were only too ready to oblige—so ready, in fact, that some kittens ironically found them a nuisance and an obstacle to their indepedence. When Papu or Nan or Kerompa could not find any kittens to nurse, they would go in search of them, calling frantically as they trotted around the building and attempting to retrieve kittens even when the latter were intent on play. Papu in particular became very concerned whenever she saw kittens scrambling around at the tops of ladders or the garden trees, and would pull and drag them, despite their complaints, down to the ground—only to watch them leap back up again the moment she released them. On more than one occasion during such endeavors, I saw her drop a kitten by accident. At such times the kittens could no doubt have wished their foster mothers elsewhere, but when they had played their hearts out, they would readily come running in response to any calls promising motherly affection and a drop of milk.

During that summer and autumn we witnessed the emergence of a few more foster mothers. With so many kittens in the cattery being brought up together in pooled litters by two or more mother cats, and these kittens then going on to receive further affection from a number of foster mothers, it is hardly surprising that such a warm atmosphere existed in general among the inhabitants, barring the perennial rivalries among the toms. With a maze of actual and perceived blood ties joining them together, the younger cats in particular had very good reasons for behavior toward each other as members of the same family, even if, in the last analysis, it was only the walls of the building that held them together as a single unit.

11

IMMIGRANTS, EPIDEMICS AND A TORTOISESHELL TOM

Among the cats in the Animal Kingdom who had not been born there, there was not one for whom a single yen had been outlaid in purchase. Whether we wanted them or not, with frightening regularity cats would find their way into our hands, bringing us both pleasure and problems aplenty.

Most of the older cats, such as Mo, Ee, Uko, Aya, Ganko and Chibi, had taken up residence with the Kingdom's official blessing, upon requests from their owners, but as the population grew, it became the policy to refuse any offers of dogs and cats. The number of requests, however, grew in direct proportion to the increasing exposure that the Kingdom was getting on television from 1980, despite our insistence that the TV companies concerned make it clear in their broadcasts that the Kingdom was not, and had never been intended to be, a refuge for the nation's unwanted dogs and cats. Especially during summer, we would get calls from throughout the country, sometimes two or three a day, asking us to accept litter after litter of puppies and kittens. Hating having to refuse, everyone developed a phobia about picking up the phone during these months, particularly as some callers would, upon being refused, dissolve into tears, or even occasionally become quite unpleasant. Thankfully, however, most callers accepted with polite resignation the reasons we gave for turning them down.

It was sad irony that, while refusing the animals of these people, most of whom, incidentally, also offered money to cover the cost of caring for them, we had little choice but to take in those animals dumped unannounced on our doorstep. Occasionally they arrived by train, usually in a sorry state and invariably with the name and address of the sender proving to be false. More often, they would be found abandoned outside the gate of the Kingdom in cardboard boxes, despite the sign that we had pinned there asking people to refrain from doing so. For this reason, any strange cars stopping momentarily outside the gate were viewed by everyone with great suspicion. I witnessed an actual dumping only once, and by the time I reached the gate, the car was already disappearing down the hill, leaving a cloud of dust in its wake which regrettably obscured its license number. I was left shaking my fist and vainly screaming a long list of expletives, both Japanese and English, which do not bear repeating here. What else can one do by way of venting one's anger at such behavior—apart from later imagining at one's leisure the kinds of punishment one would like to mete out to any culprit one was lucky enough to catch?

My favorite fantasy was simple enough: dumping the culprits for a few days on Mutsu-san's former abode, the uninhabited island of Kembogi, to give them a taste of the cold, hunger and loneliness that await most abandoned animals. This was, in my opinion, a far more fitting punishment than that prescribed by Japanese law—a maximum fine of thirty thousand yen (about two hundred dollars), hardly a sum likely to discourage a crime which is anyway one of the easiest in the world to get away with. The fact that animals abandoned outside the Kingdom gate were taken in and often found new homes in no way justified the act, which was plainly and simply one of cowardice and moral irresponsibility.

The rogue who narrowly missed getting an earful of my curses and delivery into the hands of the local police had left a box containing two very gentle adult cats, one of them a beautiful tabby-pointed male, the other a tortoiseshell and white female. There was some writing scribbled on the box which read: "Our names are Petoru (male) and Jutoru (female). We are one and a half years old, and our mother is a Himalayan. Our owner can't care for us anymore, so please give us a home."

Petoru and Jutoru were fortunately the

only adult cats to arrive in such a way during my tenure as keeper of the cats. Kittens could often be found owners, but very few people were ready to accept adults. Moreover, kittens had no problems adjusting to life in the cattery, but we learned from Petoru and Jutoru just how difficult it was for an adult to settle into the place. The problem lay not with the other cats, who showed curiosity but no aggression toward the newcomers, but with Petoru and Jutoru themselves. They reacted with predictable fear upon first being placed in the cattery. Jutoru just crouched quietly and let herself be inspected, but Petoru would not let any cats near him, hissing and flashing his claws at any who came within range.

Both cats spent their first night crouched on the concrete floor behind one of the sawdust tubs. Luckily it was summer, but even so, the cattery must have seemed a very cold, inhospitable and smelly place to them. As the days passed they began to relax and partake in meals of their own accord. They never did, however, look completely at home and happy in the cattery, even after a year had passed. Petoru in particular was a loner who shunned the company of other cats. I have already described how he insisted on eating on his own, and how he rarely instigated duels with other toms. He was no weakling, but he seemed totally uninterested in becoming a full member of the hierarchy. Even Jutoru, who was the most gentle of creatures, failed to develop any strong attachment to any cat other than her brother.

Both of them seemed unwilling and unable to become friends with their fellow inhabitants, but their affection for each other was strong and lasting. The sight of them munching together at a tin of cat food, or nestled next to each other on the window shelf of the far shed, at an hour of the day when most other cats were snoozing in sunnier locations, was both touching and sad. Although Petoru came to avail himself of other females when they were in heat (invariably after most other toms had bored of them), he would return to Jutoru's side after his passion was spent.

Petoru would not let any cat near him, except his sister Jutoru.

Petoru comforts Jutoru on top of the shutter.

He showed his attachment to her in a dramatic way whenever she climbed onto the top of the shutter in the far shed, a location from which she had great trouble in descending. Sometimes at meals, when all the other cats were busy eating, I would hear her plaintive calls from the far shed and, going to her rescue, would find Petoru up there with her. As soon as I had climbed up and lifted her down, Petoru would leap down of his own accord, which seemed proof enough to me that he had only gone up there in the first place to soothe Jutoru in her dilemma. If I needed any more convincing, I got it one day when Petoru actually came running to me upon hearing me enter the cattery. By the way he fussed around my legs and then went trotting off in the direction of the far shed, meowing repeatedly, I could not help feeling that he was begging me to follow him. When I did, sure enough, there was Jutoru once more stranded at the top of the shutter. After this incident, I could no longer watch old *Lassie* reruns on Japanese television with the same cynicism.

Jutoru and Petoru, in their relationships with each other on the one hand, and with the other cats on the other, seemed to epitomize both the social and the antisocial sides of the nature of domestic cats. While they seemed to remain out of choice on the fringe of "society," such as it existed in the cattery, their affection for each other was a telling example of the kind of warm and lasting relationships that can exist between individuals who have been brought up together. The presence of cats like them only added to the scientific interest of the cattery, and I would like to have been able to introduce more adults of varied origins and upbringings for this reason. From a management point of view, of course, an open-door policy was impossible. If we had accepted all offers of cats, we would have been swamped in no time.

What worried us far more than this, however, was the introduction of illness from outside. Until Ganko's arrival, there had been, almost miraculously, no fleas in the Kingdom. They were soon endemic. Although periodic powdering controlled them, they were never totally eradicated. After this experience we were very careful with newcomers, keeping them in quarantine for a couple of weeks before introducing them to the cattery, and giving them a thorough inspection for fleas, mites, worms and so on.

A case of cat flu.

No matter how careful one is, however, risks of infection, particularly with viral and bacterial diseases, can never be totally erased. Jutoru and Petoru had looked perfectly healthy, but ten days after introducing them to the cattery in the summer of 1981, one cat after another came down with flu. The first signs were runny noses and sneezing, accompanied by mild fever and loss of appetite. Within a day or so, however, many cats' nasal passages were blocked, and their eyes regularly glued up with discharges from infected lachrymal glands. Some showed ulceration on their tongues. Many of those afflicted also discharged copious amounts of smelly saliva which coated and matted the fur of their faces and forequarters and made them look very sorry sights. The flu virus itself is not so dangerous, but secondary bacterial infection often sets in and results in pneumonia. Moreover, even once over the preliminary fever, with their nasal passages blocked, afflicted cats lose their sense of smell and with it all desire to eat or even drink, and so death can often occur from dehydration and starvation.

Thankfully, only about fifty cats were affected. We closed the door between the two sheds and isolated the sick cats in the far shed. For two anxiety-filled weeks we fought the disease, concocting the tastiest food possible into which we mixed antibiotics to combat secondary infection. We did what we could to keep the cats clean, and their eyes and noses unblocked. Those cats most seriously afflicted and not eating or drinking at all were given twice-daily subcutaneous injections of lactated ringer solution with added vitamins, and intramuscular antibiotic injections. We managed to feed some with formula milk using stomach tubes, a practice which even the sickest cat would object to violently. With at times thirty cats requiring this kind of treatment twice a day, we were kept very busy—and gratefully so, for it always felt better to be doing something, particularly at the beginning of the outbreak when the number of afflicted cats was growing daily and our spirits were at a very low ebb. When concentrating on injecting a 100 cc liquid meal into a struggling, squirming cat, one had little time to entertain the dark thoughts which constantly invaded one's mind in any spare moment when all chores and treatment had been completed for the day.

Despite our efforts, two young toms died, but one by one the others gradually showed signs of improvement, allowing us to begin to relax our guard and once more breathe easily. Although many of the afflicted would no doubt have pulled through on their own, we could credit ourselves with saving the lives of about twenty whose condition had been very serious. Since the majority of the cats had been born in the Kingdom, and since this was the first attack of cat flu in the place, I don't know why so many cats escaped infection. Immunity to the two varieties of flu virus is not supposed to be passed on from mothers to kittens, but the evidence from the cattery suggests that a good many cats possessed a natural immunity to the virus. Strangely, however, some cats suffered while their siblings showed no signs of illness. Whatever the explanation, we were only too happy that the outbreak was limited and resulted in so few deaths. That the kittens born that year to Pe, Plum, Chestnut, Toco and Chi all escaped infection seemed little short of miraculous, even taking into account their isolation in the maternity rooms and the care we took to limit our contact with them.

Although it is impossible to say for sure, the timing of the outbreak pointed to Petoru and Jutoru, who remained unaffected, as the most likely source of infection. We had no way of knowing if they had survived a flu attack at some time in their lives prior to coming to the Kingdom, but if they had, they would have been carriers of the virus.

The next summer found us doing battle with another outbreak, two weeks after two kittens had been picked up outside the gate. Luckily we had been able to find homes for these kittens within a few days, and had kept them in isolation in the meantime; but here again, the timing of their arrival and the outbreak of flu seemed more than coincidental. The number of cats affected was about the same as the previous year, but the fact that among the sufferers were some who had also caught the first outbreak suggested that the 1982 virus was of a different strain. Moreover, although it produced much the same symptoms as those of the previous year's bug, the 1982 strain proved to be much more persistent and damaging. Despite being better prepared to deal with the situation, we lost a total of fourteen cats that year, most of them youngsters of one or two years—Bu, Fumi, Harpo, Nougat, Cinnamon, Olive, Rabbit, Mac the table tennis star, and so on, all playmates of ours who had been born and had grown up under our care. Although more than thirty other cats, including Ganko, Hi, Nibu, Panda, Kenjiro and Kenbo, responded to treatment and gave us something to cheer about, those were gloomy days. Every time I buried one more victim, I swore that I would never let another strange cat near the cattery. Even if it was not certain that abandoned cats were the source of infection, I was in no mood to take chances again, feeling that to do so would amount to betrayal of the cats already in my care. Any abandoned cats who could not be found homes immediately would, I decided, have to be put to sleep, although I hated the thought of doing this. While I was aware that an animal put to sleep with anesthetic dies painlessly, it is no fun for the person concerned to plunge the needle in, knowing that he or she is putting an end to the life of a perfectly healthy living creature. From that time on, I prayed more fervently than ever that no more cats would find their way into our hands, and at long last my prayers would appear to have been answered, for the two kittens were the last feline immigrants to the Kingdom during my tenure.

Had the first outbreak of flu occurred a couple of months earlier in 1981, I would have been forced to think twice about accepting Skipper into the ranks of the Kingdom cats. As it was, I was not aware, as perhaps I should have been, of the risk I was taking, and leaped at the offer from a lady living two hundred miles away near Sapporo—for Skipper, according to this lady, was nothing other than an example of that genetic anomaly, the tortoiseshell tom.

I could hardly believe my luck! Tortoiseshell toms are not as rare as many people suppose; maybe one in every one or two hundred tortoiseshells is a tomcat, but at the same time they cannot be encountered every day of the week. I for one had never set my eyes on such a creature. Until recently in Japan, they were particularly prized by boat owners and skippers who would outlay considerable sums to purchase them as ship cats, considering that they were invested with the power to protect vessels from storms. Although superstitions concerning their magical powers are no longer so widespread, there are still many people who will pay handsomely for one of them.

The tortoiseshell tom has for obvious reasons long been a popular subject for genetic research, but as far as I could gather from the material available to me at the time, almost nothing had been written about its behavior. This struck me as strange, since much of an animal's behavior is genetically controlled, and genetic anomalies accordingly often display anomalous behavior. Did the lack of reference to the behavior of tortoiseshell toms imply that they behave just like normal tomcats, or had their behavior simply been ignored as a subject for study? Whatever the reasons for this omission, if it indeed existed, I was interested in seeing with my own eyes how a tortoise-

"We've got a strange one here!" Skipper's first hour in the cattery.

shell tom would behave in the cattery, although I was aware that the behavior of a single example would hardly provide grounds for any firm conclusions to be drawn.

The first thing we did upon releasing Skipper from the sturdy wooden box in which he had arrived was to inspect his hindquarters.

Although we did not really doubt the word of his owner, it was a relief to confirm with our own eyes that this part of his anatomy indeed qualified him for the label "tomcat." However, in other respects, quite apart from his coloration, not only myself but almost everyone who had gathered in the living room to get a look at the strange newcomer remarked upon his distinctly untomlike appearance. According to his owner, Skipper was one year and three months old, which meant that he still had a little room for growth. There were a few toms in the cattery who at a similar age had been slighter in build, but even they had shown the heavier-set jowls by which anyone reasonably familiar with cats can distinguish at a glance the average tom from the average female. In short, not only in his coloration but also in his head shape and general body conformation, Skipper was far more reminiscent of a female than a tom.

His tail was long and straight, traditionally an unusual characteristic in Japanese cats, as I have mentioned before. I was pleased to note that he was a tricolor—that is, a tortoiseshell and white. The coat of plain tortoiseshells usually presents a rather nondescript, brindled appearance, being composed of very small patches of orange and black fur. For reasons not yet well understood, the presence of white patches has the peculiar but pleasing effect of reducing the mixing of black and orange. The more extensive the white area, the more clearly the remaining tortoiseshell areas become segregated into larger patches of either orange or black. Thanks to the large areas of white on his belly, legs, flanks and throat, Skipper's tortoiseshell areas were relatively clearly divided into orange and black patches, and he was as a result a handsome (or perhaps, in view of his physique, I should say "pretty") specimen. Skipper's mother, with whom he had spent his life up to that point, was also a long-tailed tortie and white. Since she had been free to wander out, Skipper's father could have been one of any number of toms who apparently inhabited the neighborhood.

Skipper passed his first two weeks in the Kingdom in the living room. He was an exceptionally friendly cat, and the speed with which he settled down to his new life there was almost disconcerting. He showed none of the initial fear that most cats display when finding themselves suddenly in strange surroundings. Apart from this, the only unusual aspect of his behavior noticed during those first two weeks was his almost religious use of the lavatory. I cannot think of one of the cattery toms who at the same age had not been fervently and liberally spraying his surroundings. According to Skipper's owner, he had shown the same absence of spraying in her home. He had also apparently shown no sexual interest in his mother when she had come into heat a month previously.

Despite Skipper's calm response to the living room, we fully expected a more typical reaction of fear from him when we introduced him to the cattery. Here again, however, he baffled us with his nonchalance. He neither uttered a single hiss nor raised a single paw in defense against the horde of cats who gathered to investigate him. Far from flattening his ears with fear, he pricked them forward in intense curiosity; not, however, for the cats, but for his surroundings. After exchanging nose greetings with a few of the cats, he set off almost immediately to explore the building. For all the notice he took of them while wandering, the party of cats that trailed him might just as well not have existed. He was clearly excited by his new home, for he spent a few hours sniffing around it before finally settling down alone to snooze, long after most of the other cats had lost interest in him.

There were, however, a few young toms who continued to pester him, attempting to mount him whenever he stopped long enough to allow them to do so. While cats are almost color-blind, the toms would nevertheless have been able to recognize his patchwork pigmentation in monochrome, and thus had every excuse for mistaking him visually for a female. Moreover, he still smelled different from a long-term inhabitant, and young toms in particular considered any strange-smelling cat to be a suitable target for their sexual advances. For this reason I took no special notice of such behavior at that time.

I was forced to revise my opinion, however, when Skipper continued to prove attractive to a few toms, and was willing to accept their advances, even some weeks after he had taken up residence in the building. Although he did not go out of his way to grab the attention of other toms, he seemed to have no objection to being mounted, and on some occasions even looked as though he positively enjoyed the experience. He would never rush off when a tom, giving up the impossible act in frustration, dismounted him, but usually remained in a crouched position, his rump slightly raised, for a few seconds after. Twice

Skipper appeared to like being mounted.

at such moments I witnessed him actually look around and beckon his suitor with little chirruping calls of the same kind that a frustrated female will often utter. At one stage of that summer, he was being mounted regularly enough for the nape of his neck to become calloused and stippled, much like that of many females during estrus.

Skipper was certainly proving to be a very strange character. While there was no reason to connect his nonchalance toward new surroundings and strange cats with his being a tortoiseshell tom, upon observing his willingness to be mounted by other toms, I could not help wondering whether this behavior was a manifestation of a possible feminizing influence caused by the presence of an extra X chromosome. Other behavior that he did or did not display over the next few months only reinforced such wild speculation—for example, his attitude to the females.

By then, almost all the females had undergone fallopian tube cuts, as a result of which there were always some in heat from January through to September. The conditions were more ideal than ever for any tomcat worthy of the label to prove his manhood, and most of the toms were busy both night and day doing just that. Skipper, however, was a notable exception. Not only did he continue to show no signs of spraying activity, but he also showed not the slightest romantic interest in females, even though he was several months past the age when healthy young toms have normally chalked up their first sexual conquests.

On the contrary, when the mood took him, he could be downright aggressive toward the females. This was really extraordinary behavior. Normally tomcats fight with other toms, and females with females, although much less frequently. Instances of toms fighting with females were very rare indeed in the cattery, and no single tom ever consistently instigated quarrels with females—that is, until Skipper! Although he by no means picked quarrels every day of the week, on every occasion that I witnessed, he was the perpetrator of the altercation, and his targets were always females.

The fact that his opponents were usually different on each occasion indicates that he did not bear a grudge against any one female in particular, but was just letting off steam, so to speak, as females, too, were prone to do occasionally. His opponents reacted in much the same way as females always did when suddenly threatened by another female, with a look as much of surprise as of fear. They seemed to find his attacks as unexpected and inexplicable as I did. I never once saw any cat, male or female, start a fight with Skipper—almost as if they, too, found him a somewhat ambiguous character, difficult to pin down as either a male or a female.

It was after the beginning of the next year that Skipper began to behave in a more orthodox manner. He ceased picking on females, instead showing some sexual interest in them, and actually mating with a few. He also began to spray at the same time, but in neither activity was he nearly as impassioned as the average tom. When one of Chestnut's female offspring from the previous year showed signs of coming into heat for the first time in early summer, instead of operating on her I immediately isolated her with Skipper. He was not a very fervent suitor, but dutifully mated with her three times, after which I kept her in the main house for observation. She turned out not to be pregnant, dashing my hopes on that occasion that Skipper might be a fertile tortoiseshell tom. However, the fact that she did not conceive was by no means enough evidence to rule either way on the question of Skipper's fertility.

I was planning to pair him with other still sexually intact females born the previous year when they came into heat, but before any of them did, tragedy struck in the form of the second flu outbreak. Skipper was one of the victims, even though he had been left unscathed by the previous year's epidemic. He died in the middle of the outbreak, despite our efforts to save him, leaving a lot of unanswered questions concerning his behavior. Was he merely a slow starter where sex and spraying were concerned, as sometimes even genetically normal toms are? Or was his lack of manly passion a direct effect of the kind of anomalous sex chromosome constitution found in other tortoiseshell toms? Were his tendency to aggression toward females and his willingness to be mounted by other toms during the first few months also effects of the same genetic anomaly?

Skipper had spent little over a year in the cattery, but I think that his behavior in that time was interesting enough to indicate that investigation not only of the genetic constitution of tortoiseshell toms but also of their behavior would be highly rewarding.

12

TOBO AND BLOSSOM

Among the hundred and twenty-odd cats that we cared for in the Animal Kingdom, there are some who, because of their individuality, or because of special circumstances that linked them to us, will be remembered vividly and dearly by my wife and myself, probably for as long as we live: "Lord" Mo, "Bovver Boy" Waru, those two fearless rat catchers Marshmallow and Kenbo, "Sexpert" Wanchan, Ping-Pong star Mac, our foster kittens Otochantachi, Chiyoko's favorite female Pe, tortoiseshell tom Skipper, those two shy and affectionate siblings Petoru and Jutoru. However, no account of our involvement with the cats would be complete for us without special mention of two particular feline friends, Tobo and Blossom.

These two cats, both females, now exist only in our memories and in the photographs I took of them. Tobo died of cancer during the early summer of 1981, and Blossom in the spring of the following year, even before reaching the age of one, again of cancer. Although there were many excellent mother cats in the cattery, Tobo seemed to stand out even among them as the personification of feline motherhood. And Blossom, who was reared alone from the second day of her life by Chiyoko, perhaps brought us more laughter while she lived, and more sorrow in her death, than any other kitten we had had the pleasure of knowing. Although Tobo and Blossom spent only three weeks of their lives in each other's company, for special reasons, their names will always remain linked in our memories.

Tobo was the first cat that I met when I arrived in the Kingdom as a foreigner in the summer of 1976. There were other cats inhabiting the living room at that time, but when I was ushered into that room for the first time, my eyes immediately fell on her. She was a fine-looking chocolate-pointed Siamese in all respects but one: her face was, to put it bluntly, grotesque. Her upper lip appeared to be almost totally eaten away, leaving her incisors and canine teeth permanently on view—just the kind of face, at first glance, which would have had the casting director of a horror movie drooling at the mouth and leaping with joy. Noticing me stare at Tobo, Mutsu-san picked her up in his arms and, before even introducing me to everyone else in the room, briefly related her history in very rusty English which was nevertheless surprisingly strong on vocabulary.

Tobo had been picked up two years previously during the winter of 1974, on the day after a blizzard. She was already an adult at that time, at a guess about two or three years old. She had been found plodding very wearily along a snowbound lane near the Kingdom. Indeed, she got her name from her appearance when picked up, for the Japanese translation of "plodding wearily" is *tobo tobo aruku*. She was clearly a house cat, fond of people, for she had approached when called and allowed herself to be picked up. She had been very thin, and probably not far from death by starvation and exposure. There had been a big open sore on her upper lip and smaller ones on her legs. The tips of her ears appeared frostbitten. Since cats have excellent, almost supernatural "homing" abilities and rarely get lost of their own accord, it was assumed that Tobo had either escaped, say from a car while being transported, or more likely had been abandoned by her owners, maybe outside the Kingdom gate. All inquiries in the neighborhood as to her origins had drawn a blank, and so she became a permanent member of the Kingdom household. She was from the start apparently a very affectionate cat, and soon became a firm favorite of the Kingdom's human inhabitants—all the more so because of the circumstances of her arrival, since animals which have been picked up on the brink of death invariably secure a special place in the hearts of those who

Tobo.

have rescued them and brought them back to health.

If indeed Tobo had been abandoned, Mutsu-san guessed that the big sore on her upper lip could have been one of the reasons why. Another vet residing in the Kingdom at that time had diagnosed it to be what is known as a rodent ulcer, a misleading name since rodents have nothing to do with the causation, which is internal. Since all available medicines had failed to have any effect on the ulcer, the only treatment left had been excision of the affected tissue—and that, in short, is how Tobo came to lose most of her upper lip. The operation had left her marked for life, but had been successful insofar as the ulcer had not recurred, as they are sometimes prone to do, even after excision.

Thanks to Mutsu-san's efforts with his English, Tobo was the first animal in the Kingdom whose background I got to know in any detail. Within a couple of days her face no longer appeared grotesque to me; in fact, she even began to appear beautiful, the better acquainted I became with her, and her presence in the living room was always comforting. At that period in her life, she was a veritable kitten factory, and always seemed to be pregnant or nursing a litter of kittens. And she would continue nursing any kittens for whom we had not found homes, almost up to the time of birth of her next litter.

One would think that even the most diligent of mothers would tire of such a regimen, but not Tobo! I remember an occasion in the spring of 1977 when, for a change, she was kittenless. Ishikawa-san had brought into the living room a newborn fox cub that he was hand-rearing. The cub's cries immediately attracted Tobo, who, after sniffing the cub, latched her teeth gently but firmly onto its nape and tried with all her might to pull it from Ishikawa-san's grip. She so pestered him to hand over the cub that he soon removed himself and the cub from the room, after which Tobo spent a good half hour meowing, sniffing and

Tobo with her 1980 foster kittens.

scratching at the door through which they had disappeared.

For some reason, perhaps because her womb had done so much work up until then, Tobo seemed to have exhausted her reproductive potential by 1979. The two kittens that she gave birth to, a couple of months before she was transferred to the cattery along with the other cats, were her last. In the cattery, she became pregnant twice, but both times these pregnancies ended in miscarriages. Instead of rearing her own kittens, she devoted herself to the care of others. She was, indeed, the first foster mother to emerge in the cattery.

In 1979, one of her last two offspring, Papu, gave birth to her first litter, and when we released Papu and her kittens from the maternity ward, Tobo immediately switched her attention from other, larger kittens whom she had been fussing over to these, her grandchildren. Although they were the smallest kittens in the cattery at the time, I could not help feeling that Tobo was drawn to them not only for their size but also because she knew them to be the offspring of her daughter. These suspicions were only reinforced by my observations over the following two years on the degree of coop-

Kittens play with each other . . .
. . . with a beetle . . .
. . . with a twig . . .
. . . with a paper fish.

Playing with the corpse of a baby rat.

(above left)
A wrestling match in the grass.

Blossom plays inside a paper bag.

Genta, a deer with a broken leg, provided a warm sleeping place for Blossom in the fold of his legs.

During her short life, Blossom loved all creatures, human and animal, and had a succession of playmates. Malibu, a cross between a wild boar and a domestic sow, was a somewhat unwilling punching bag.

(left)
Tobo and Blossom.

Skipper, that strange rarity, a tortoiseshell tom.

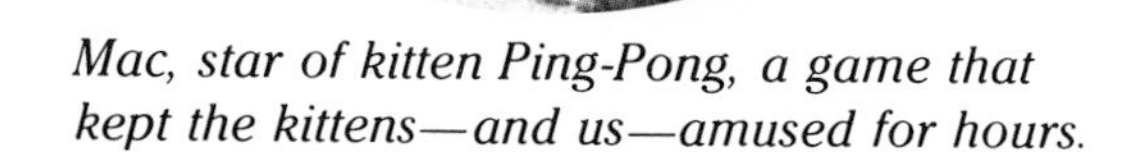

Mac, star of kitten Ping-Pong, a game that kept the kittens—and us—amused for hours.

Petoru the loner.

Petoru and Jutoru, the first adult newcomers to the cattery.

eration shown by closely related females. Significantly, Papu, who so resembled Tobo in body size, conformation and color, also appeared to have inherited her mother's exceptional maternal instincts, and was the most diligent of the foster mothers who subsequently emerged in the cattery.

In 1980, for the first time in almost two years, Tobo got the chance to nurse newborn kittens again when three were rejected soon after birth by their mother. Chiyoko had once more volunteered herself as foster mother, but this time she decided to enlist some help, since we were both hectically busy at the time. Accordingly, Tobo was brought from the cattery to attend to all the needs of the kittens other than their milk. In view of her proven enjoyment of rearing kittens, she was the ideal choice, and she did not disappoint us. She immediately moved the kittens from the box in which Chiyoko was keeping them to what she considered to be a safer location, deep inside Chiyoko's wardrobe. She took over almost completely, reducing Chiyoko, albeit gratefully, to the role of mere "wet nurse." After a week, however, Tobo seemed increasingly keen to take on this role as well, daily objecting more strongly to Chiyoko removing the kittens to feed them, trying to pull them from Chiyoko's hands back into their nest.

We were not yet aware at that time of the ability, mentioned in Chapter 10, of some females to produce milk for kittens that they had adopted, even without themselves giving birth, and so when she first observed the kittens suckling on Tobo's teats, Chiyoko thought that they were getting little more than emotional sustenance. However, when the kittens began accepting less rather than more formula milk with every passing day, and when, despite this, their little bellies proved to be strangely full even four or five hours after their previous feed, she was baffled. Just on the off-chance, she inspected Tobo's chest closely, and there she found the answer to the mystery. Even if not to the degree of a genuine mother with newborn kittens, Tobo's teats were undeniably swollen, and when squeezed, sure enough, tiny droplets of rich, white milk oozed forth. Tobo thus became the first cat in the Kingdom to display this phenomenon, at least to our knowledge. Within two weeks of handing the kittens over to Tobo, there was no longer any need for Chiyoko to feed them. From that time until they were weaned, Tobo raised the three kittens solely on the strength of her own teats.

Tobo had looked the picture of health and happiness with her foster kittens during that spring and summer, but by the beginning of the next year, we noticed the appearance of a peculiar skin ulcer on the rear of her left hind leg. Iwase-san considered the ulcer to be one of the same class of skin ulcers to which rodent ulcers belong. The prescribed treatment was the same as before: try medication, and if this fails, operate. And once more, all the medicines we tried were ineffective. Although the ulcer did not appear to bother Tobo much, there was a chance of it spreading and so we decided to operate. In the end, we operated twice, but to no avail. No matter how boldly we excised, the new tissue that grew around the scar was as ulcerated as the old.

There seemed little point in further operation; having spent over two months, barring mealtimes, with a mask around her neck to prevent her licking the affected area, Tobo had suffered enough. Once the operation scar had healed as much as it ever would, we removed the mask for good and Tobo got down to living a less restricted life with her ulcer. She had passed most of the winter in the living room and, as she seemed very happy there, we let her remain. Everyone spoiled her frightfully, and she got as much as she wanted of all kinds of tasty morsels that her fellow cats in the cattery rarely got a chance to sample. The ulcer, which we had reverted to treating with medicines, neither improved nor worsened. Gradually, however, as the days became steadily warmer, Tobo's appetite declined and her eyes occasionally bore a faraway look. She took less and less interest in the life of the living room, and spent most of every day sitting quietly in patches of sunlight.

It was one such day when Chiyoko suddenly embraced Tobo, saying, "I know what'll put some life into you, old girl—a kitten! And I've got just the one for you."

With that, she set Tobo back down on the sofa, disappeared from the room and returned a few seconds later with Blossom cupped in her hands.

Blossom was a tiny cream-colored kitten, just three weeks old—the one that we had removed from Plum on the day after her birth, for fear that she would be crushed in the nest her mother was sharing with Pe, Chestnut and their kittens. She was not only an exceptionally small but also a very strange-looking kitten at first, with only her head appearing to be of normal size. When she was about a week old, for some inexplicable reason, the fur on the crown of this head began to fall out, only adding to our concern that there was something congenitally wrong with her. Oblivious of our worried gazes, however, she would grip the

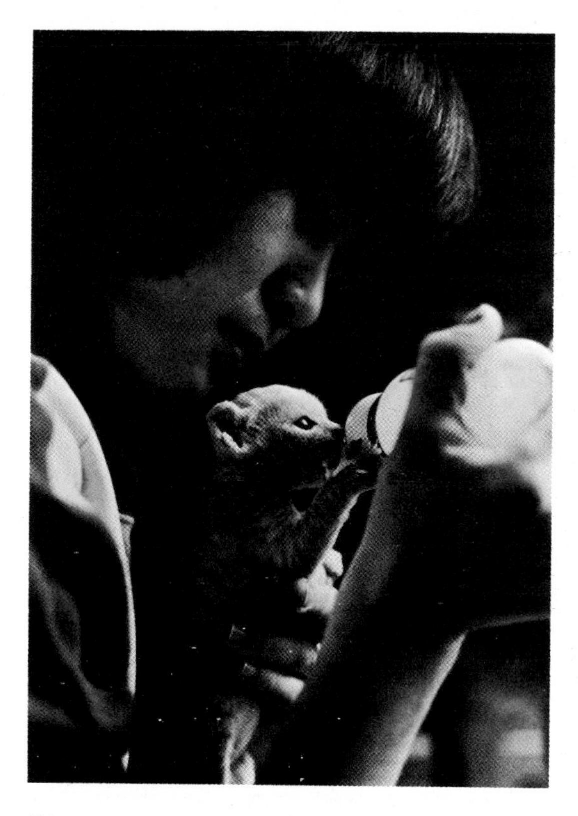

Blossom would grip the rubber bulb of the feeding bottle with both forepaws.

rubber bulb of the feeding bottle teat with both forepaws (after she had given up hope of getting mother's milk—once again Chiyoko had had to struggle for two days before persuading her to accept a feeding bottle), and then suck for all she was worth, like a perpetual motion machine, until her belly was literally full to bursting. She clearly had every intention of surviving, even if we entertained doubts about her ability to do so. Chiyoko had called her "Blossom" because of her color and because the wild cherries in our remote region had finally come into bloom around the time of Blossom's birth.

Chiyoko was very attached to Blossom, the more so because of all the time and worry she had expended in the first few days of caring for her, and so it was with some surprise that I observed her willingness to hand the kitten over to Tobo. It was, however, a great idea, judging by Tobo's reaction. The minute she saw the kitten, her ears pricked forward, her eyes lit up and her lethargy left her. In no time at all, she had once more relegated Chiyoko to the status of wet nurse, and for the following three weeks she hardly let Blossom out of her sight. Blossom could not venture even a yard from her nest without Tobo leaping out to retrieve her. Apart from these times, however, Blossom seemed only too happy in Tobo's company, and could be found curled up in Tobo's embrace whenever she was not intent on play.

The union was tragically short-lived, however. Early one morning I came into the living room to find Tobo not with Blossom, but sitting quietly on top of the refrigerator. Still suspecting nothing, I lowered her, wished her good morning and placed her with Blossom, who had been asleep on the sofa. Tobo sniffed the kitten cursorily and then climbed onto the back of the sofa and leaped once more for the top of the refrigerator. The distance was about a yard, diagonally upward. It was a leap that she had always managed with consummate ease, but this time only her forefeet got as far as the edge of the refrigerator and she fell to the floor.

There was something about this failure and the clumsiness of her landing which alerted my suspicions. I took her in my arms and began to feel her body here and there. It was then that I discovered that, although not visibly so, the lymph glands under her chin were definitely swollen. I immediately checked those in her groin, and found these, too, to be swollen. Blast it! I thought. Whatever it was that was ailing her, this was not a good sign. No doubt someone would have discovered it earlier had Tobo not been so wrapped up with Blossom. She had been eating, even if not voraciously, and nothing in her behavior over the previous three weeks had led anyone to suspect that there was anything wrong with her.

When Iwase-san came in from the stables for breakfast, I asked him to take a look at Tobo. He confirmed that her lymph glands were indeed swollen, and immediately took a blood specimen. He then prepared and stained a blood smear, and looked at it under Mutsu-san's microscope.

"Bad news," he said, staring down the eyepiece. "It looks like there are far more white blood cells here than there should be, and what's more, there are very few normal ones among them. It's almost one hundred percent certain that Tobo's problem is lymphosarcoma—in other words, cancer—and I think you know what that means."

Yes, I did. I had done enough reading to know that any treatment would be merely palliative, there being no cure for such a condition. Moreover, these cancers are almost always caused by a virus known as feline leukemia virus. Although this virus is not highly infectious, any cat showing symptoms of the disease should at least be kept out of contact with others. The most sensible measure, from the point of view of preventing both transmission to other cats and any further suffering on the part of the patient itself is of course euthanasia, and this is what Iwase-san recommended.

Chiyoko removed Tobo to our room, where she settled, still as a rock, on our radiator, not asleep but with her eyes shut. She showed not the slightest interest in food or anything else, and we did not have to be mind readers to see that she was suffering. Neither Chiyoko nor I saw much point in prolonging her ordeal and so, later that morning, we put her to sleep. She passed away quietly in Chiyoko's arms.

As Iwase-san had requested, we let him peform a postmortem immediately afterward. The results were completely as he had expected, there being not one area of lymphatic tissue in her body unaffected.

With a kitten to care for, Tobo was once more her old self.

According to Iwase-san, the cancer could well have taken hold some months before, and was probably one reason why the ulcer on her leg had shown little response to treatment.

"Judging from this evidence, it's a miracle that Tobo didn't die much earlier. Although there's no way of telling for sure, I reckon she survived the last two or three weeks in such apparent health because of Blossom's presence," he said as he pointed out one affected area after another.

Chiyoko and I were of the same opinion. During the previous three weeks, Tobo had been her old self; perhaps not as active, but as attentive and happy as she had always been when in the company of kittens. Blossom had given her a new, albeit brief, lease of life. We could only hope that the kitten would not be made to pay for it as a result. This thought particularly bothered Chiyoko, who poured her heart out to me that night.

"I had a dream about Tobo dying, and so decided the next day to give her Blossom," she admitted tearfully. It was the first time that she had touched on the subject, and her talk of a dream came as quite a surprise. She is a very practical, down-to-earth person who rarely delves into her dreams, and was not given, either before or after this occasion, to experiencing premonitions. "I didn't really believe the dream, but Tobo had seemed a little somber. She had always been happiest with kittens to look after, and so giving Blossom to her somehow seemed the right thing to do. If only I had known Tobo was suffering from such a disease! I should have got Iwase-san to look at her then, instead of doing what I did."

Under the circumstances, it was hardly surprising that Blossom's death the next spring from the same disease hit Chiyoko the hardest. Although she had been with Tobo for only three weeks, it appeared that she had been unlucky enough to be infected. Thankfully, however, she was remarkably healthy for most of her short life, and left all of us with many amusing and touching memories. She was such a favorite with everyone that she was kept in the living room and, lacking any other feline playmates, she displayed an extraordinary interest in human affairs. I have never known any kitten blessed with the company of other kittens to show much interest in the repetitive batting of a human eyelid or protrusion of a human tongue, but Blossom used to play for minutes on end with my tongue and eyelids.

She also quickly got on friendly terms with other animals who passed time in the living room. Her first nonhuman playmate was Malibu, who, for the sake of convenience, I shall call a boarlet since she resembled her father, a tame wild boar, more than her mother, our resident domestic sow. Both boarlets and piglets, when they can get their minds off their stomachs, play delightfully and are great fun to have around. Malibu was no exception, and that is why she was in the living room. She was, however, no match for Blossom and had to be endlessly long-suffering. She would just get settled down on the sofa with a full belly when Blossom would scramble up the back of the sofa and leap on her from above. The kitten would then start to play with Malibu's ears, causing her to flick them constantly, this flicking only adding to the fun. Finally, after driving the poor boarlet almost to distraction and making her get up and down from the sofa time and time again, Blossom would snuggle down beside her, using her nose as a pillow.

Blossom also got on famously with a Maltese puppy who spent a few days in the Kingdom for the shooting of a television commercial with the other animals, but perhaps her greatest companion of that period was Genta, a Hokkaido deer fawn who had been reared from birth by one of the girls. Genta had ventured too near to the rear of the horses on one of his daily walks in the pasture, and one of them had kicked out in fright, snapping one of his hind legs. This was immediately set in plaster, and Genta spent the next month, until the break mended, in the living room, forcing Malibu, who was a little too big and boisterous under the circumstances, back outside with his brothers and sisters.

Blossom and her playmates: playing dead . . .

Blossom immediately switched her attention to the newcomer, seeming to find him an even more ideal playmate and pillow than Malibu. Genta always stood up very suddenly, so as to avoid putting weight on his broken leg, and because he did so even when Blossom was asleep between his legs, basking in the warmth of his chest, she was trodden upon on more than a few occasions by his sharp hooves. Every time this happened, she let out such a scream that we feared that this time for sure she had been seriously injured. Luckily, however, she never suffered more than momentary pain, and this was not sufficient

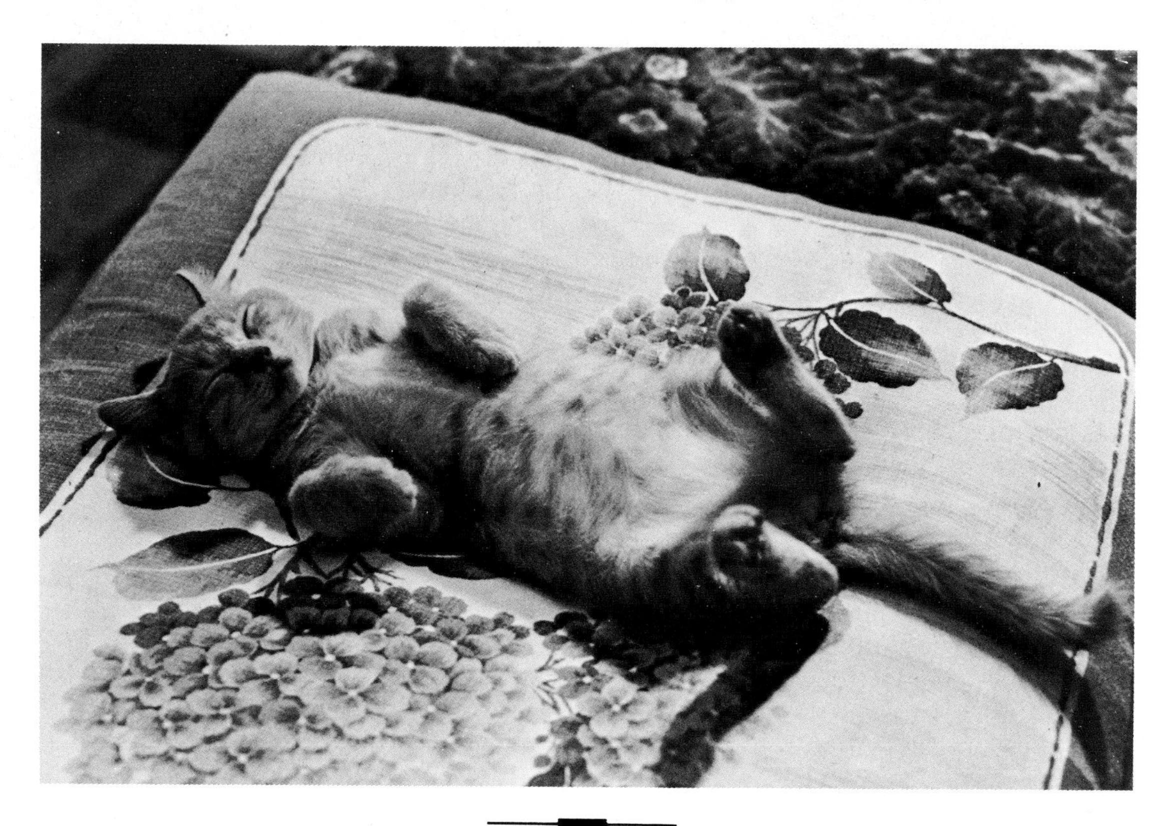

with Genta the deer . . .

to discourage her from returning to the fold of Genta's legs the moment he had settled down again—whereupon she would frequently receive, with apparent enjoyment, a thorough lick-down from the fawn. The play of these two infants was no less touching, Genta trying to butt Blossom, while she lurked under chairs and poked her paws out at his nose and feet, then scurrying off with Genta in pursuit around the long dining table.

Not surprisingly, Blossom came to consider anything on four legs to be a friend and source of amusement, and so whenever we had reason to bring any of our dogs into the living room, or at least those with murder records, we had to keep Blossom elsewhere. Only when the dog concerned was Boss, our Saint Bernard, was Blossom allowed to indulge her playful spirits, since cats scared Boss stiff! Blossom spent a good half an hour tormenting poor Boss by play-

With Malibu the boarlet . . .

. . .with her last playmate,
Rocky, the black bear cub.

ing with the latter's tail; this was the only interesting part of her anatomy available, since Boss spent the entire time keeping her face as far away from Blossom as she possibly could. I believe that even if an elephant had suddenly barged into the room, Blossom would have responded in the same manner. She was innocent and fun-loving to the point of recklessness.

Blossom's illness brought grief to the whole household. She lost her appetite and her joy for life and just began to waste away. And when cancer was diagnosed, we once more had no choice but to part with a dear companion. She was buried next to Tobo's grave on the hill outside the Kingdom fence, facing the Pacific Ocean. In addition to Tobo and Mo's sister May, many of the Kingdom's other deceased inhabitants were also buried there—Mutsu-san's legendary brown bear Donbe, his two Akita hounds Guru and Ken, the stallion Tani, the three-legged wild boar Bui, to name but a few. For Blossom, who seemed to love all creatures great and small, and had little idea exactly what kind of creature she was herself, there could not have been a more ideal resting place.

The domestic cat still retains much of the solitary nature inherited from his wild ancestors.

13

SAYONARA

Chiyoko and I left the Animal Kingdom on May 10, 1983, catching the early morning train so as to avoid drawn-out good-byes, something that both of us dreaded. Chiyoko, who was famous for her proneness to tears in emotional situations, laughed and bantered constantly through breakfast, through luggage-loading and through our very brief farewells with the animals, so much so that anyone who did not know her would no doubt have considered her to be a very coldhearted person indeed. She was, however, fooling only herself with her display of jollity, and no one in the Kingdom would have been surprised to know that she finally allowed the floodgates to open the moment that the train pulled out of the station—to my great relief, since I was having trouble holding back my own tears.

The timing of our departure was decided largely by the discovery, in January of that year, that Chiyoko was pregnant. Mutsu-san had no objections to any of his staff (provided they were married!) raising their children in the Kingdom, as proven by the fact that Ishikawa-san and Hiroko were already the parents of two. However, both Chiyoko and I had imposed on Mutsu-san's benevolence for over six years—a good deal longer than anyone but the Ishikawas. We sensed that we had gained as much from, and given as much to, the Kingdom as we were ever likely to, and with the prospect of parenthood looming on the horizon, we felt the need for an environment which offered more independence and privacy than were possible in a tiny self-contained community like the Kingdom, even if this meant sacrificing for the time being the kind of involvement with animals and nature to which we had grown accustomed.

Never having intended the Kingdom to be a permanent home for the young people whom he took in, and being well aware of the difficult aspects of life there, Mutsu-san gave his blessing to our decision, although he was more than a little concerned about my ability to make ends meet in the "outside world." I have to admit that, outside that particular neck of the Hokkaido countryside, Japan was still very much a foreign country to me, and our future was very uncertain; but of far more immediate concern to Chiyoko and myself was the future of the animals we would be leaving behind.

We were not so conceited as to think that they would miss us nearly as much as we would miss them, or that those who stepped into our shoes would be any less proficient at taking care of them. Nevertheless, we had shared so many years with the cats and dogs that we could not help feeling guilty about just walking out of their lives. The fact that we would have been faced with the same situation whenever we had chosen to leave did little to alleviate our feelings. The least that we could do was remain in the Kingdom until suitable replacements had been found and taught the ropes of caring for the cats and dogs, and so we left the date of our departure open.

Toward the end of March, Keiko, an ex-nursery school teacher, and Yuko, a trained veterinary assistant, arrived, just as Chiyoko had done six years earlier, from a spell on a dairy farm in western Hokkaido. Both were twenty-five years old, very strong, capable and easygoing girls who showed a natural aptitude and boundless delight in dealing with animals. We let them take over one task after another until, within two weeks, we found ourselves virtually redundant in everything but an advisory capacity. Consequently, we were able to meet the day of our departure with much lighter hearts than when we first announced our decision.

I had embarked upon the cattery project with varied aims in mind: to give the cats occupying the main house more space and more freedom, to photograph their behavior in a way that would be both scientifically meaningful and pleasing to the eye,

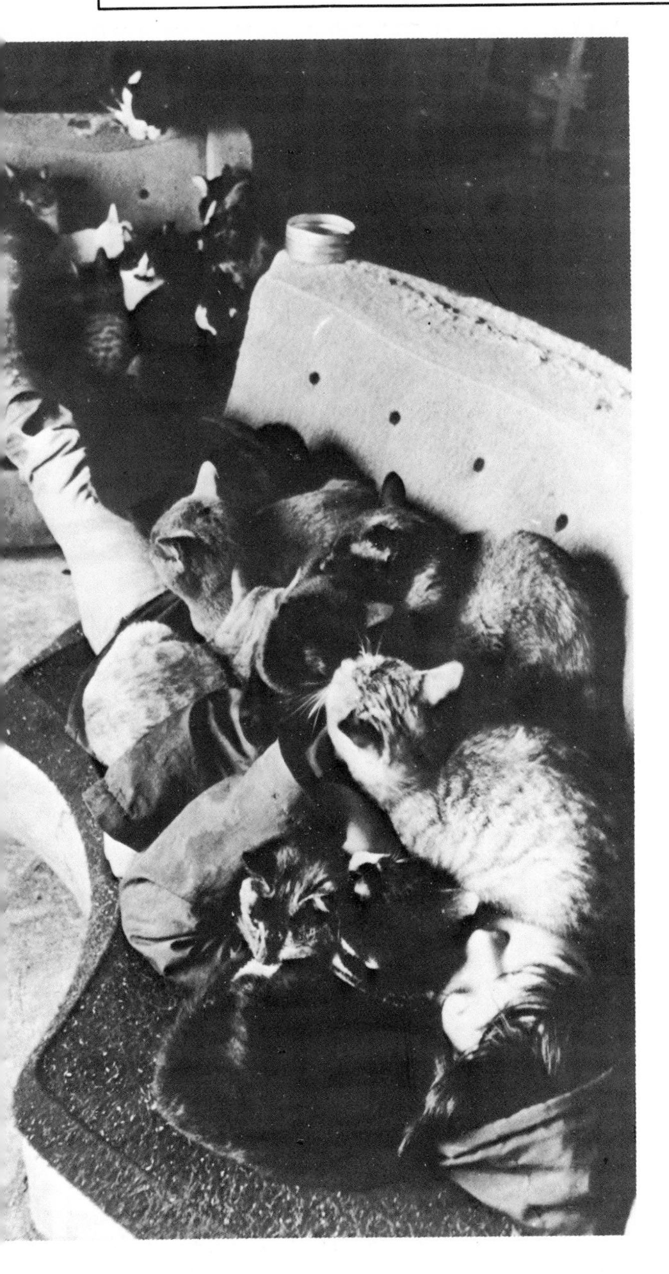

For Chiyoko, an electric blanket would be a poor substitute.

and to investigate their social relationships and hopefully draw some conclusions concerning feline social behavior in general from the events witnessed in the somewhat special situation of the cattery.

The first of these aims was of course achieved on day one of the cattery, and received a further boost with the addition of the garden. One can do little more than speculate on whether the cats actually enjoyed life there more than in their old quarters, but if they could have made their opinions known, I think that the large majority of those who knew both worlds would have opted for the cattery. On the other hand, Kagetora and his kind would no doubt have been hard pushed to find one good word to say for the place!

The photographs are really a matter for readers, not myself, to pass judgment on, although if the reaction of readers of the Japanese edition is anything to go by, I can claim some success. I was overjoyed by the number of appreciative letters I received. There would appear to be a demand among both cat lovers and those interested in animals in general for a more respectful approach to the cat, one which treats it as more than a charming fluffy toy or mantelpiece ornament; rather as an animal whose behavior is every bit as fascinating and worthy of serious photographic representation as any other animal which inhabits this earth. This is just one attempt at such an approach, the photographs portraying the natural behavior of the cats in this particular environment.

And what conclusions concerning the

social behavior of cats can be drawn from the findings of the cattery? I think that in general the behavior of the cattery inhabitants demonstrated both the solidity of the solitary nature inherited from the wild ancestors of the domestic cat and the thickness of the veneer of sociability laid down on that solitary base through the process of domestication. The insistence of most of the cats most of the time on "doing their own thing" (admittedly within the limits allowed by their environment); the great predominance of individual over group activity; the practice of a "live and let live" principle; the disinterest in either wielding or accepting authority, and the resulting lack of pervasiveness of the influence of the tomcat hierarchy over most aspects of life; all these show the cat to be still very much the captive of its wild past as one of this planet's most solitary and independent animals.

This need hardly surprise us in view of the shortness of the cat's history of domestication (at the most adventurous of guesses based on existing evidence, no more than about six or seven thousand years—a mere jot in time compared with the few million years that its wild relatives have inhabited the earth) and the fact that, even after domestication, little more was required of the cat than to do what came most naturally to it as a wild animal: hunting of small prey such as mice and rats, an activity which is necessarily solitary.

Nevertheless, when compared with its wild relatives, the domestic cat is a very sociable creature. In the cattery, most of the cats got on with each other remarkably well. The lack of friction between the females and the warm cooperation that many of them displayed in kitten-rearing were especially impressive, but no less so was the tolerance shown toward each other by the toms, despite their frequent duels which, in my view, tended to make them look far less sympathetic characters than they actually were. The most likely ancestor of the domestic cat, the African wildcat (*Felis lybica*), is significantly one of the most easily tamed of the various species of wild cat, but when compared with the domestic cat, it still possesses a pretty intractable nature. A male and female African wildcat (and male/female pairs of most other wild cat species) can live together in captivity fairly peacefully, but we could not expect over a hundred of them to coexist with anything like the harmony displayed by the occupants of the cattery.

Insofar as it is the only one of man's domestic animals that can boast truly solitary roots, the cat is indeed an enigma, but almost certainly for the same reason, it is a very successful enigma. While it is able to form warm and lasting relationships with humans and members of its own kind (and with other animals also), it has no overwhelming need for company, and reserves the right to choose the degree of its involvement with others according to its mood or situation. For every house cat that is willing to share its home with others, one could probably find one who will absolutely refuse to do so. And yet many of the latter kind, if they are free to wander outside, probably consort as friends with other cats in the neighborhood at certain hours of the night and at certain locations, and not only for the purpose of mating, but simply out of a desire for some degree of social contact. Similarly, there are strays or semi-strays who lead largely solitary lives, and others who live as members of small colonies on farms or in factories and warehouses. Unless a cat has been born into such a colony, becoming a member of one might be tricky, but not impossible if the events described in *The Curious Cat* are anything to go by.

The inhabitants of the cattery had little choice in the matter of living companions, and there was great pressure on them to socialize to the limits of their abilities. However, even under these circumstances they demonstrated the same flexibility in various aspects of their behavior toward each other. On cold winter nights it was unusual to find any cats except Kagetora and other, more temporary pariahs sleeping alone. Most of them piled, one on top of another, into the two winter beds, each of which could hold more than fifty cats at a push. Even those who preferred less crowded conditions could be found sleeping in groups of three or four in various other strawlined boxes I had prepared. They slept together, of course, mainly to keep warm, as proven by the fact that, on warm summer days, they would snooze scattered through the building, either singly or in small groups, this choice being dictated far more by each individual's mood or desire for company than by temperature. In other words, if one could disregard climate, the cats appeared equally happy either sleeping alone or sharing a bed with their fellows.

Not so the dogs outside, whose sleeping arrangements were far more stereotyped. At times they seemed almost pathetically dependent on each other when compared with the cats, who were far more flexible or, if you like, opportunistic, and gave the impression of being able to take or leave their "society" as situation or mood dictated. Clearly, cats can get on with each other, and often do to great mutual advantage and with obvious pleasure, but they can also get on without each other. It is no doubt such flexibility which has enabled the domestic cat to adapt with such success to all manner of environments, both man-made and natural.

It is Christmas Eve of 1984 as I end this last chapter. We have just watched a special program filmed in the Kingdom, and were delighted to see, among all the familiar faces, those of Mo, Ee, Uko, Aya, Nibu, Wanchan and a host of other old feline friends, all of them with thick winter coats shining and well groomed, looking the picture of health. A notable absence from the lineup was Pe, who, sadly, died last spring. Thankfully, however, she has been the only casualty. Although a few more abandoned kittens have been taken in since we left, there have been no more epidemics.

In her latest letter, received a couple of days ago, Keiko informed us that the snows had arrived much earlier than usual, forcing her to take down the net and close the

Ee and a female companion snooze happily together.

garden in a hurry in the middle of this month. She reported that Mo was still the undisputed number one tom, and still not showing his age in the slightest. Here's wishing him and all his fellow cats in the Animal Kingdom happiness and the best of health for many years to come.